U0163178

中西餐合璧菜文化丛书

Chinese and Western
Fusion Cuisine
China and Canada

赵建民　高速建　总主编
［加］岳昌龙　主编

中西餐合璧菜
中国和加拿大

中国轻工业出版社

图书在版编目（CIP）数据

中西餐合璧菜：中国和加拿大 / 赵建民，高速建总主编；（加）岳昌龙主编. — 北京：中国轻工业出版社，2024.4

ISBN 978-7-5184-4895-1

Ⅰ. ①中… Ⅱ. ①赵… ②高… ③岳… Ⅲ. ①中式菜肴—烹饪 ②西式菜肴—烹饪 Ⅳ. ①TS972.11

中国国家版本馆CIP数据核字（2024）第051544号

责任编辑：贺晓琴　　责任终审：李建华　　设计制作：锋尚设计
策划编辑：史祖福　贺晓琴　　责任校对：朱　慧　朱燕春　　责任监印：张　可

出版发行：中国轻工业出版社（北京鲁谷东街5号，邮编：100040）
印　　刷：艺堂印刷（天津）有限公司
经　　销：各地新华书店
版　　次：2024年4月第1版第1次印刷
开　　本：889×1194　1/16　印张：9.5
字　　数：218千字
书　　号：ISBN 978-7-5184-4895-1　定价：138.00元
邮购电话：010-85119873
发行电话：010-85119832　010-85119912
网　　址：http://www.chlip.com.cn
Email：club@chlip.com.cn

231563K9X101ZBW

作者简介

赵建民　Jianmin Zhao

山东旅游职业学院副教授

山东鲁菜文化博物馆名誉馆长

世界中餐业联合会专家委员会研究员

西安欧亚学院中国国际食学研究所受聘专家

山东省文化和旅游厅非物质文化遗产评审专家库成员

高速建　Sujian Gao

四创大世界基尼斯纪录——最薄的手拉活海参

世界中华厨师联合会国际名厨专委会主席

世界中餐业联合会职业技能竞赛国际评委

第二届中国烹饪世界大赛状元

俄罗斯联邦总厨协会顾问

国家文旅部2018欢乐中国年行走的年夜饭（俄罗斯）代表团执行团长

资深级注册中国烹饪大师

国家发明专利“海参膜字画”发明人

岳昌龙　Changlong Yue

加拿大使里刻国际咨询有限公司餐饮顾问

世界中华厨师联合会国家烹饪队专委会主席兼国家烹饪队总教练

加拿大中华烹饪协会副会长

世界厨师联合会注册认证国际（A类）评委

加拿大国家高级烹饪职称技能鉴定考评师

加拿大注册高级烹调师

英国葡萄酒与烈酒教育基金会认证中级葡萄酒品酒师

中式烹调高级技师

中西餐合璧菜文化丛书编委会

本书编委会

名誉主任： 黑伟钰

主　　任： 唐习鹏

副 主 任： 方　浩　郭洪义　刘晓南

主　　编： [加] 岳昌龙

副 主 编： [加] Serge Belair　李　辉　李建辉

编 撰 人： 赵建民　高速建　[加] 岳昌龙

英文翻译： [加] 岳昌龙　[加] Joel Kwek

菜品制作： [加] 岳昌龙　常龙山　李红良　刘恩林　姜洪伟　刘军朋　周伟琪

摄　　影： [加] 岳昌龙

联合出品单位：

世界中华厨师联合会中西融合菜专委会＆国菜国汤研究院

加拿大联邦烹饪协会

加拿大北艾尔伯塔理工学院

加拿大使里刻国际咨询有限公司

加拿大粤之味餐饮集团

大国味道美食大健康产业研究院

北京神州网云教育科技有限公司

北京屈浩烹饪服务职业技能培训学校

指导单位：

世界中餐业联合会国际中餐名厨专业委员会

序一

在“中西餐合璧菜文化丛书”系列之《中西餐合璧菜：中国和加拿大》一书付梓之际，受世界中餐业联合会国际中餐名厨专业委员会与“中西餐合璧菜文化丛书”编委会的邀请，为该书写 篇序。从而认真拜读了《中西餐合璧菜：中国和加拿大》一书，深感内容很好。此次正式出版，可喜可贺！世界中餐业联合会的职责之一就是推动中西餐的融合发展，《中西餐合璧菜：中国和加拿大》一书的编写与出版，本身就是一件值得大力倡导和赞誉的事情。

我国倡导的“携手构建人类命运共同体，为世界谋大同”的价值观，是中华民族对世界贡献的中国智慧。在这个人类命运共同体的共建中，包括人类饮食文明的交流融通，而弘扬中餐饮食文化、传承中餐烹饪技艺、讲好中餐饮食故事、推动中餐产业发展，是世界中餐业联合会的历史使命。中餐的全球化发展是每一个餐饮人的心愿；通过餐饮人的努力，让中餐技艺与西餐技艺融合发展，将是推动中餐走向世界的有效途径之一。

早在我国改革开放之初，高速建大师就提出了“中西餐合璧菜”的理念，经过几十年的理论研究与经营实践，中西餐合璧菜形成了丰富多彩的、中西烹饪技艺融合的新菜式，为中餐烹饪技艺的创新发展发挥了积极的作用。但是，迄今为止，以中西餐合璧菜为主要内容的专业著作在我国似乎仍然是一个空白。今天，《中西餐合璧菜：中国和加拿大》的出版，将会填补这一空白，其深远的意义不言而喻。

在此，谨代表世界中餐业联合会向为《中西餐合璧菜：中国和加拿大》一书的出版付出劳动与心血的指导者、编写者、出版者表示衷心的感谢！并希望“中西餐合璧菜文化丛书”能够硕果累累，为推动中西饮食文化的交流发挥应有的作用，为中西餐烹饪技艺的融合发展作出新的贡献。

是为序。

世界中餐业联合会会长

邢颖

邢颖

2024年1月于北京

序二

正所谓“南橘北枳”，由于地理位置、气候的不同，每个地方都有自己特有的饮食文化，根据食材、习俗、口味及烹饪风格的不同，形成了中国八大菜肴风味体系，主食上有“南米北面”的区别，口味上有“南甜、北咸、东酸、西辣”的特色。

即使是同一个国家的不同地区，饮食文化也存在显著的差异，更不用说中西方之间的饮食文化差异了。正因为如此，如果我们将博大精深的中华饮食文化原封不动地引入西方餐饮市场，是绝不可能被西方主流饮食文化全盘接受的。这也是高速建大师的理念。

攻读东方美食学、中国烹饪大师高速建的高徒岳昌龙先生长期生活在加拿大，一直从事和餐饮产业有关的职业，他在实践中积累经验，在学习中探索融合，将理论和实践相结合并付诸行动，就地取材、创新产品，于是便有了今天的“中西餐合璧菜文化丛书”系列之《中西餐合璧菜：中国和加拿大》。

此书介绍了用加拿大本地食材结合中华料理理念中“食养”的精华，加上符合当地人口味的创新佐料，将中西饮食文化相融合，顺应消费趋势，以食载文、以文兴商，坚守中华饮食文化的传承，与时俱进，在升级打造符合新时代口味的美味佳肴的同时又彰显了色、香、味、形、器协调一致的中华传统文化底蕴，是值得学习和收藏的好著作。

在此向赵建民、高速建、岳昌龙以及精心编制《中西餐合璧菜：中国和加拿大》一书的出版工作者表示衷心的感谢，并期待将来有更多的好作品给我们的生活添彩加色!

世界中餐业联合会副会长
加拿大餐饮总会会长
加拿大和平饭店集团总裁
加拿大和平传媒董事长
加拿大和平学院创办人
黄汝週

黄汝週

2023年9月于加拿大温哥华

序三

The Canadian Culinary Federation delighted to know that Mr. Changlong Yue, under the guidance of his mentor, Mr. Sujian Gao, has compiled a series of "Chinese and Western Fusion Cuisine" books, and that the book *Chinese and Western Fusion Cuisine: China and Canada* is about to be published. On behalf of the Canadian Culinary Federation, I would like to congratulate the publication of this book and hope that it will play a positive role in promoting the exchange of Chinese and western food culture and culinary skills. China is a country with a long history in food culture and their culinary skills are extremely developed. Whether in China or in western countries such as America and Canada, the fusion development of Chinese and western culinary skills has become a new trend in the world's catering industry! We believe that the publication of *Chinese and Western Fusion Cuisine: China and Canada* will definitely play an important role in promoting the exchange of Chinese and western food culture, promoting the fusion development of Chinese and western culinary skills and making new contributions to it.

Sincerely
Chef Ryan Marquis
National President, Canadian Culinary Federation

October 2023 in Toronto, Canada

序四

烹饪是科学，是艺术，在现代思想文化交融下，人们对食文化不局限于单一食的思维方式，他们积极去寻找不同领域食的享乐。

在食文化的传播中，人们也不限于地域，积极开展思想文化交流与学习。随着时代的发展，今天的厨师面临一个重大的研究课题，即要求现代厨师在继续教育和学习方面，发挥更加专业的力量，将不同的菜系和菜品融为一体，具备开拓中西菜品融合的新思维。通过烹饪的实践与探索，提升厨师对中西菜品融合新思维的认识，为食文化的发展开辟更广阔的道路，并传承食文化的故事。

在人类发展史上，我们今天的厨师，不单是一个厨艺家，而且是一个烹饪思想家。

在中西饮食文化融合发展的前提下，将中西烹饪技术互相融合创新的新菜式，编写成为《中西餐合璧菜：中国和加拿大》一书，将有助于更多的人更加全面地了解中西菜品，在文化领域达到一个创造性的高度，作出一个现代厨师对社会应有的贡献。

加拿大中华烹饪协会会长

廖志文

Michael liao

2023年8月10日

前言

按理，前面已经有了世界中餐业联合会会长邢颖先生、加拿大餐饮总会会长黄汝週先生、加拿大联邦烹饪协会会长瑞安·马奎斯和加拿大中华烹饪协会会长廖志文先生的序，足以为《中西餐合璧菜：中国和加拿大》一书的出版增色。但踌躇再三，自认为还需要再写一个前言，对“中西餐合璧菜文化丛书”总主编之一高速建先生与《中西餐合璧菜：中国和加拿大》一书的主编岳昌龙先生略加介绍。

高速建先生是我国烹饪行业内的著名大师，烹饪技艺高超，业内成绩斐然。在中国烹饪圈内的大多数人都知道，高速建是我国烹饪泰斗王义均的弟子，得到了王老烹饪技艺的真传，尤其在烹饪和研究海参菜肴方面，成绩卓著，素以“手拉活海参”和烹制海参菜肴而闻名。然而，却很少有人知道高速建大师早在30多年前就开始了对于“中西餐合璧菜”的临灶实践与理论研究。差不多可以认为，对中西餐合璧菜进行理论研究的，高速建在我国厨师界堪称第一人。

高速建早在1989年在山东省烹饪学会会刊《烹饪者之友》杂志上，就发表了《中西餐合璧菜肴》的短文，是对研制中西餐合璧菜肴的经验和技艺介绍。之后几年，高速建依然对中西餐合璧菜的理论研究矢志不渝，并于1993年在《烹饪者之友》杂志第二期上发表了《融中西烹技，制人间至味》的文章。在该文章中，高速建先生对中西餐合璧菜烹饪技艺的渊源、必要性、概念、特点，以及中西餐合璧菜的应用方法进行了初步的探讨，引领了国内烹饪行业中西餐合璧菜理论研究的风气之先。

下面就把高速建先生于1993年发表的文章全部转录如下，作为本书“前言”的主干内容。

融中西烹技，制人间至味

我的《中西餐合璧菜肴》一文在贵刊1989年第三期上发表以后，得到了同行和前辈们的肯定。鲁菜大师、第二届全国烹饪技术比赛评委郭延祥认为，中西餐合璧菜是适应时代发展的创举，是鲁菜及中国烹饪向科学化、营养化高层次发展的一个重要方面。青岛名厨尹顺章在贵刊1990年第一期撰写的《继承改革创新——振兴鲁菜之我见》一文中写道：“《烹饪者之友》载文，研究中西餐合璧菜肴，可谓是鲁菜的大胆创新”。这对笔者无疑是极大的鼓舞和鞭策。为了深入研究中西餐合璧菜，使之上升到理性认识，笔者把将近十年来探索、研究和积累的实践经验，加以概括、总结，整理成下面粗浅的一孔之见，盼望得到烹饪界的专家和学者的批评指正。

一、中西餐合璧菜的概念和特点

所谓中西餐合璧菜，就是中餐菜和西餐菜相互渗透、交叉、有机结合而制成的菜肴。

中西餐合璧菜具有它自己显著的特点：它既保持了中餐菜的味道，又具有西餐菜的风格；既能照顾到外宾的饮食习惯，同时又考虑到内宾的饮食风俗。在制作上，它融汇中西餐菜肴之长，吸取各自的精华，使之在色、香、味、形、营养等方面更臻完美，从而形成了中西餐合璧菜肴的特有风味。

二、研究中西餐合璧菜的必要性

中国菜肴以色、香、味、形、器俱佳而驰名中外。但是，世界烹饪之纷繁复杂、丰富多彩，是由众多的各国烹饪所构成的，中国烹饪是世界烹饪的组成部分，各国烹饪都有其特色，各有千秋。例如，许多国家已将科学手段应用到烹饪之中（例如法国的“真空烹饪法”等），特别重视营养、卫生对人的作用等，这恰巧是中国烹饪的弱点。还有如风味化学，这门在20世纪50年代就已诞生的学科，我国至今仍很薄弱，甚至还未应用到烹饪研究之中。以法、英、美、意、俄等国为代表的西餐菜肴具有主料突出、形色美观、口味鲜美多样、富于营养、品种繁多等特点。那些口味各异的“沙司”调味品，具有开胃、清口、帮助消化的特点，突出酸、甜、香、辣的各种“色拉”等，都是值得我们学习和借鉴的。尽管中餐和西餐各自有不同的理论体系和烹调方法，但是，由于两者都是为了供给人体需要的各种营养素，达到保证人体正常发育和健康的目的。这样，中西餐之间就有交叉点，就有共同规律。因此使中餐和西餐相互交流融汇，努力发挥各自的优势，探索其共同规律，找到两者的交叉点，研制中西餐合璧菜，已成必然之趋势。中西餐合璧，在烹饪领域开辟了一条新路子，它解决了中餐菜和西餐菜各自的局限性和不足，为创制出独具特色、科学营养的“人间至味”，提供了新的广阔天地。

三、中西餐合璧菜的渊源及发展过程

中西餐相互交流、结合的历史，可以追溯到汉代。汉武帝时期，朝廷派官多次出使西域各国，与中亚、西亚以至欧洲各国之间，广泛进行了经济、文化、饮食的交流。汉明帝时期，随着佛教在我国盛行，佛家吃斋，不茹荤腥，对我国饮食产生了广泛的影响，外国的食物如胡麻（芝麻）、胡瓜（黄瓜）、胡荽（元荽）、胡椒等传入我国。外国的烹饪技法如胡羹、胡炮等也融入了我国的食谱，从而推动了中国菜肴的革新，这是历史上中西餐的第一次人文交流，从此便开始了中西餐的相互吏融、结合。以后各代，随着陆上、海上交通的发达，进一步扩大了与世界各国的文化、饮食交流。南宋杭州有款名菜叫冻波斯姜豉，即是传自伊朗。到了清朝，我国对外交流范围更加广泛，中外饮食交流遍及全球。光绪年间，由我国人自己开设的以营利为目的的“西餐馆”“面包房”等，开始陆续出现在上海、北京、天津等城市，甚至西太后举行国宴招待外国使臣，有时也用西餐。这期间，像“沙司”“色拉”“吐司”之类的外来烹饪术语也开始进入我国。经过一段时期的经营，我国的西餐业有了很大的发展，并形成了许多风味不同的帮派。有的经营正宗的欧美菜，有的经营俄式大菜，有的经营中国味的西餐（也称番菜），还有的经营家庭“番菜”等。其中，中国味的西餐和家庭“番菜”即是中西餐合璧菜。这一时期的中西餐的相互结合较前代有了很大的进步。

纵观从汉代到1949年前中西餐相结合的历史，中西餐合璧菜经过从无到有、不断地发展，已经成为中国菜肴的一部分。随着我国对外开放和旅游事业的不断发展，我国与世界各国之间多层次、多渠道、多形式的友好往来日益频繁。笔者认为，在新的历史条件下，我们应当学习各国烹饪研究的成果，研究世界上的菜系和风味流派、世界烹饪的发展趋势、各国菜系交叉的风味等，应用现代科学

（如生物学、化学、营养卫生学、美学等）分析研究中西餐菜肴的色、香、味、形、营养等方面，以及制作过程的变化规律。有目的、自觉地、科学地在更大的范围内使中西餐相互融汇、结合，研制出更多的中西餐合璧菜肴。实践证明，中西餐合璧菜的确可以起到中国菜和西餐菜所起不到的作用和效果，是中餐菜和西餐菜无法代替的。因此，可以预见，不久的将来，中西餐合璧菜将以其独特的风味和绚丽的风貌在我国菜肴中独树一帜。

四、中西餐合璧菜的结合方式

中西餐合璧菜由于结合的方式不同，可分为以下几种类型：

第一种类型：在配料上中西餐相互组合，融为一体。例如："面包炒肉片"这款中西餐合璧菜肴，是我国传统菜"滑炒肉片"配上西餐中的面包片创制而成的。"滑炒肉片"鲜嫩适口，面包片过油后清香脆酥，将这两者科学组合，两种不同风味集于一盘，其味之美，不言而喻，颇受中外宾客的青睐。

第二种类型：中西餐原料互相补充。如：用西餐菜点中常用的结力冻代替中餐菜中的琼脂冻，用中餐中的海螺代替法国名菜"焗蜗牛"中的蜗牛等。

第三种类型：中西餐调味品互相丰富。例如：中餐菜调以咖喱沙司、番茄沙司、法国沙司、辣酱油等西餐调味品；西餐中的烤火鸡等菜肴，调以中餐中的怪味、麻辣等复合味汁等，使之产生独特风味。

第四种类型：在刀工切配、拼摆造型、食品雕刻等技法上，中西餐相互借鉴。比如西餐借鉴中餐细腻的刀法等，中餐借鉴西餐中的挤马乃司、土豆泥花边和黄油、奶油裱花等。

第五种类型：烹调方法上中西餐相互引用。譬如中餐中有的品种采用西餐中的红酒焖、奶汁烩等技法，西餐中的烤鸡采用类似广东盐焗鸡的盐焗技艺等。

第六种类型：取长补短，集中西餐之优点有机结合。即将中西餐中的若干款菜肴的长处综合在一起，设计新式品。比如"空心黄油虾球"（制作方法发表在1985年第七期《中国烹饪》上）这款中西餐结合菜，即是分别取中餐中的"空心虾球"和西餐中的"黄油虾卷"的优点，有机结合创制而成的。

这篇2500多字的文章，在今天看来也许并不那么引人入胜，但却几乎是我国第一篇专门研究中西餐合璧菜肴的理论文章，今天读来仍然有振聋发聩的感觉。因为，这是一位潜心于中国烹饪技艺研究的中国烹饪大师在30年前发表的真知灼见，今天依然令人肃然起敬。

一直以来，高速建先生就在酝酿出版一本"中西餐合璧菜"烹饪技艺的专著，30年后的今天，终于在其得意弟子的共同努力下，得偿夙愿，也算是了却了心中的一桩牵挂。

或许，有众多的序言在前，《中西餐合璧菜：中国和加拿大》一书的"前言"属于一篇标新立异的文章，因为算不上是一篇序言，所以以前言的形式出现。但我们认为，把30多年前引领中西餐合璧菜研究的理论文章，放在这篇"前言"里，远超出了一般前言的意义。

或许，这本身就是一个创意之举。

目 录 CONTENTS

Part 1 第一部分

正餐菜 | Main Dishes

Part 2 第二部分

杂碎菜 | Chop Suey Dishes

特别荣誉菜
Special Honor Dishes

Part

1

第一部分

正餐菜

Main Dishes

“正餐菜”释义

所谓“正餐菜”，是基于西餐正规的、按照程序化接待客人用餐的各种菜品。

习惯上，西餐的一个正餐的上菜程序包括：前菜（Appetizcr）、餐汤（Soup）、副菜（Side Dish）、主菜（Main Course）、沙拉（Salad）、甜品（Dessert）、饮品（Drinks）等几个部分。

西餐的正餐的组合菜品，类似中餐的一桌宴席菜肴组合，但在数量上远远不及中餐宴席。在西餐的正餐中，按照一般的理解，餐汤、副菜、主菜都属于“正餐菜”，但以“主菜”的丰盛情况代表一个正餐的水平高低。

以西餐正餐中的第三道菜“副菜”而言，其用料包括水产类、蛋类、面包类和酥盒类。鱼类、贝类及软体动物肉质鲜嫩，比较容易消化，通常在肉禽类菜肴前面上桌，作为西餐的第三道菜。常见的副菜有香煎鳕鱼、蔬菜焗青口、烤三文鱼等。

而第四道菜则是以肉禽类菜肴为主，因而被称为“主菜”。肉类菜肴的原料取自牛、羊、猪等各个部位的肉，其中最具代表性的是牛肉和羊排。禽类菜肴的原料取自鸡、鸭、鹅等，其中较为经典的当数香醇嫩滑的法式鹅肝。

本书中的“正餐菜”以“中西合璧”的风格见长，是以西餐主菜的食材，融合中餐的烹饪技法制作而成，具有中西合璧、食味交融、文化融合的特点。

一、松露加拿大西冷牛排
No.1 Truffled Canadian Striploin Steak

小记 Notes

此菜在2020年2月德国IKA奥林匹克国家厨房烹饪竞赛中获得主菜银牌。此菜选用加拿大本地食材西冷牛排作为主料，运用传统西餐牛排的烹调方法加上中式烹调技法的炸牛肉棒、炒蘑菇胡萝卜和中式焗土豆，呈现出中西餐味蕾融合的巧妙搭配。

This dish won the silver medal for the main course at the Restaurant of Nations during the Germany IKA Culinary Olympic in February 2020. The main ingredient is striploin steak, which is a local ingredient from Canada. This dish combines the traditional western style of cooking steak with Chinese cooking methods such as fried beef sticks, stir fried mushrooms and carrots, and Chinese-style baked potatoes. It presents an ingenious combination of ingredients which satisfies both western and Chinese taste buds.

主料
Main materials

加拿大西冷牛排。

Canadian beef striploin steak.

配料
Ingredients

泡芙，鲜松露薄片，菜花，胡萝卜，鸡油菇，香芹末，香菜末，青葱末，圣女果，圆葱末，姜末，蒜末，土豆，鸡蛋，淀粉，罐装鹅肝酱。

premade puff, thin sliced fresh truffles, cauliflower, carrots, chanterelle mushrooms, minced parsley, minced cilantro, minced green onions, cherry tomatoes, minced shallot, minced ginger, minced garlic, potatoes, eggs, starch, canned foie gras mousse.

调料
Seasonings

黄油，芥花油，鲜奶油，豪达奶酪，蜂蜜，百里香，黑胡椒，白胡椒粉，酱油，味极鲜酱油，大蒜粉，孜然粉，鸡粉，盐，粘肉粉，面包糠，酱汁。

butter, canola oil, fresh cream, Gouda cheese, honey, thyme, black pepper, white pepper powder, soy sauce, Weijixian soy sauce, garlic powder, cumin powder, chicken powder, salt, meat glue powder, breadcrumbs, sauce.

制作过程 Methods

1. 将整条西冷牛排修整成需要的形状，保留边角料。将修好的牛排从中间片开不要切断，再放入鲜松露薄片，均匀撒上粘肉粉，然后用保鲜膜紧紧裹住放入保鲜柜至少30分钟。取出牛排，均匀撒上盐和黑胡椒。锅中加入芥花油、黄油烧热，加入百里香，放入牛排煎至表面金黄取出，放进烤盘中，放入60℃烤箱烤25分钟取出。

   Trim the striploin steak to the required shape, keeping the trimmings. Bufferfly the steak, place the thin sliced fresh truffles into the meat, evenly sprinkle with meat glue powder, then wrap tightly with saran wrap and put it in the fridge for at least 30 minutes. Take out the steak and sprinkle evenly with salt and black pepper. Heat canola oil and butter in a pan, add thyme, sear the steak until the surface is golden brown, take it out, put it on a baking tray. Roast in a 60℃ oven for 25 minutes and remove.

2. 将取下的牛排边角料剁成肉馅，加入青葱末、圆葱末、姜末、香菜末、酱油、盐、黑胡椒、鸡粉、鸡蛋、淀粉调成馅料，做成圆柱状，裹匀蛋液，再均匀裹上香芹末和面包糠的混合物，放入油锅炸至外焦里嫩取出。

   Grind the beef trimmings into the filling, add minced green onion, minced shallot, minced ginger, minced cilantro, soy sauce, salt, black pepper, chicken powder, eggs and starch and mix well by hand. Make a cylinder shape, wrap the egg mixture evenly, coat evenly with a mixture of minced parsley and breadcrumbs, deep-fried until the outside is crispy and the inside is cooked and tender.

3. 将土豆切成薄片，加入鲜奶油、青葱末、盐、孜然粉、大蒜粉、鸡粉拌匀，将土豆片一层一层摆到不锈钢浅方盘子内，锡纸盖住放入180℃烤箱烤1.5小时取出，用另外一个大小一样的不锈钢浅盘装重物压在土豆表面，放入冰箱冷冻。待冷却后取出土豆切成长方体，上方铺上豪达奶酪片，放入60℃烤箱烤25分钟取出。

   Slice the potatoes, add fresh cream, minced green onions, salt, cumin powder, garlic powder, chicken powder and mix well. Put the potato slices layer by layer in a shallow hotel pan, cover with foil, and put in a 180℃ oven for 1.5 hours. Take it out, put another same size hotel pan with a heavy weight on top of the potatoes and put them in the fridge. After cooling, take out the potatoes and cut them into rectangle shape, put one slice of Gouda cheese on top and place in the oven at 60℃ for 25 minutes and remove.

4. 将胡萝卜、圣女果放入盐开水中煮熟去皮。锅中放入芥花油，加入青葱末、姜末、蒜末爆香，放入胡萝卜、鸡油菇和圣女果，加入盐、鸡粉、味极鲜酱油翻炒均匀。

   Add carrots and cherry tomatoes into salt water and boil and peel. Put the canola oil in the pot, swear the minced green onion, minced ginger, minced garlic until soft. Add carrots, chanterelle mushrooms, cherry tomatoes, salt, chicken powder and Weijixian soy sauce, and toss evenly.

5. 将菜花切成小块备用。锅中加入黄油、鲜奶油、盐、蜂蜜、白胡椒粉和菜花煮至菜花软糯，倒入搅拌机内搅拌成泥状，取出备用。然后将罐装鹅肝酱装入裱花袋中挤满泡芙内部。

   Cut cauliflower into small pieces. Add butter, fresh cream, salt, honey, white pepper powder and cauliflower to the pot and cook until the cauliflower is soft. Pour it into a blender and blend into a smooth puree. Take it out and set it aside. Fill the inside of the puffs with the canned foie gras mousse by using piping bag.

6. 将以上的成品摆盘，先放菜花泥，再依次放鸡油菇、胡萝卜、炸好的牛肉段、鹅肝泡芙、土豆、圣女果、牛排，最后放酱汁。

   Put cauliflower puree first, then one by one, add chanterelle mushrooms, carrots, crispy beef, foie gras puffs, potatoes, cherry tomatoes and steak. The last step is to pour in the sauce.

技术解析 Technical resolution

1. 牛排中间放入粘肉粉之后，一定要放入冷藏室中冷却30分钟以上，这样粘肉粉才能发挥它的作用。

   After putting the meat glue powder in the middle of the steak, it must be put in the fridge for at least 30 minutes, so that the meat glue powder can work.

2. 豪达奶酪不宜在超过60℃的烤箱内长时间烤制，时间过长或是温度过高豪达奶酪都无法达到要求的效果。

   Gouda cheese cannot bake in a 60℃ oven for a long time. If the time is too long or the temperature is too high, the cheese cannot achiere the desired effect.

酱汁制作
Making sauces

在平底锅中融化黄油，加入圆葱末煸香，加入牛汤大火烧开，然后改成微火熬至汤汁黏稠，加入盐、胡椒粉调味，过滤去渣成酱汁。

Melt the butter in the saucepan, and swear the minced shallot until soft, add the beef broth and reduce the broth until slightly thick, add salt and pepper powder to taste. Filter and remove residue to make sauce.

二、比目鱼和龙虾佐食白酒酱汁
No.2 Halibut and Lobster with White Wine Sauce

小记 Notes

此菜在2020年2月德国IKA奥林匹克国家厨房烹饪竞赛中获得前菜银牌。比目鱼和龙虾佐食白酒酱汁是使用加拿大本土盛产的海鲜食材，运用中西餐混合的烹调方法加工而成。

This dish won the silver medal for the appetizer at the Restaurant of Nations during the Germany IKA Culinary Olympics in February 2020. The halibut and lobster with white wine sauce is cooked by using Chinese and western cooking methods with local seafood ingredients sourced from Canada.

主料 Main materials

龙虾，扇贝肉，比目鱼。

lobster, scallops meat, halibut.

配料 Ingredients

青豆，蒜末，青苹果粒，圆葱粒，白萝卜薄片，食用明胶。

green peas, minced garlic, diced green apple, diced shallot, thinly sliced white radish, gelatin.

调料 Seasonings

黄油，猪油，鸡蛋清，苹果醋，白葡萄酒，葱姜汁，盐，白胡椒粉，龙虾汤，鲜奶油，美乃滋，绿紫菜末，脆脆豆。

butter, lard, egg white, apple cider vinegar, white wine, onion and ginger juice, salt, white pepper powder, lobster broth, fresh cream, mayonnaise, green seaweed powder, crispy beans.

制作过程 Methods

❶ 将扇贝肉加入葱姜汁、鸡蛋清、白胡椒粉、盐放入搅拌机中打成扇贝肉泥，取出一部分装进裱花袋中，挤进圆柱形模具中，用保鲜膜包住，放入53℃的蒸箱中蒸28分钟取出，将剩余的扇贝肉泥加入绿紫菜末混合均匀备用。

Place scallops meat, onion and ginger juice, egg white, white pepper powder and salt into the container of a food processor. Process until pureed. Take out a portion, put it in a piping bag and pipe into a cylinder mold, wrap it in saran wrap, and steam in a steamer at 53℃ for 28 minutes. Add green seaweed powder to the remaining scallop puree and set aside.

❷ 将龙虾尾去壳，在底部顺刀切至龙虾皮，不要切断，挤入混有绿紫菜末的扇贝肉泥，然后用保鲜膜卷成圆形，放入50℃的水槽中煮20分钟，捞出切成0.5厘米厚的圆片。

Remove the shell from the lobster tail, cut the bottom of the lobster into the lobster skin without cutting through the skin, pipe it into the scallop puree mixed with green seaweed powder, and roll it into a circle with saran wrap. Put it into a water tank at 50°C and cook for 20 minutes. Pull out and crosscut into slices approximately 0.5 cm thick.

❸ 锅中加入黄油中火加热，加入蒜末、2/3的青豆煸炒，慢慢加入鲜奶油，加入盐调味。将剩下的青豆倒入搅拌机内搅拌成泥。将青豆泥一分为二，一半加入少量的食用明胶，另一半放到60℃的保温箱。

Melt the butter in the pan, add minced garlic and 2/3 green peas and slightly toss, slowly add fresh cream, season with salt. Place the rest of the green peas into a blender and blend into a puree. Divide the puree into two equal parts. Add a small amount of edible gelatin into one part and mix well. Put the other half in a 60°C hot box.

❹ 将比目鱼宰杀去皮洗净，去边角料，切成长方形块状，用盐和白胡椒粉腌制，鱼的上面盖上切片的猪油，放入200℃烤箱烤6分钟。将边角料放入蒸箱蒸熟取出凉凉，加入青苹果粒、苹果醋、圆葱粒、美乃滋、盐搅拌均匀，做成椭圆状，用白萝卜薄片包裹四周，上面挤入加有明胶的青豆泥抹平，撒上脆脆豆。

Fillet the halibut, peel off the skin, trimmed and cut into rectangular blocks, marinated with salt and white pepper powder, covered with sliced lard, and placed in an oven at 200℃ for 6 minutes. Put the fish meat trimming into the steamer to fully cook. Take it out and chop it. Let it cool, then add diced green apple, apple cider vinegar, diced shallot, mayonnaise and salt, mix well. Make an oval shape, wrap around with the sliced white radish, pipe gelatin green peas puree on the top and add the crispy beans on top of the puree.

5. 锅中加入白葡萄酒烧至酒精挥发，加入龙虾汤、鲜奶油小火加热至汤汁黏稠，加盐调味，关火加黄油至融化，混合均匀。

   Add the white wine to the pot and reduce the wine by half. Add the lobster broth and fresh cream over the low heat until the sauce is slightly thick, season with salt, turn off the heat, add butter until melted and mix well.

6. 将椭圆的凉拌比目鱼肉放在盘子的左上角，然后顺时针依次放入烤比目鱼、扇贝肉泥，将龙虾肉片放在扇贝上面，在旁边抹上青豆泥，再撒上余下的青豆。酱汁放在中间。

   Place the oval cold halibut in the top left corner of the plate, then place the grilled halibut, scallops, and place lobster slices on top of the scallops. Spread green peas puree on the side, sprinkle with the rest of the green peas, and place the sauce in the middle.

技术解析 Technical resolution

1. 青豆泥在保存过程中温度不可以过高，否则易变色。

   The temperature of green peas puree should not be too high during storage, otherwise it will easily change color.

2. 制作龙虾、扇贝要注意温度的把控。

   When cooking lobster and scallops, pay attention to the temperature control.

3. 熬制酱汁时应小火，避免酱汁和黄油分离。

   When cooking the sauce, use low heat to avoid separation of the sauce and butter.

三、黑蒜羊肉方旦土豆
No.3 Black Garlic Lamb and Potato Fondant

小记 Notes

此菜在2020年2月德国IKA奥林匹克主厨餐桌烹饪赛中获得银牌。此菜选用羊鞍肉和羊肩肉两个部位。羊鞍肉的烹调方法运用西餐和中餐混搭而成。羊肩肉采用酱牛肉烹调理念来完成，使此菜既有西餐的灵魂又有中餐的理念。

This dish won the silver medal for the main course at the chef's Table Competition during the Germany IKA Culinary Olympics in February 2020. This dish contains lamb saddle and lamb shoulder. The lamb saddle is prepared by combining western and Chinese cooking techniques. The lamb shoulder is cooked similarly to spiced beef, which carries the spirit of Chinese cuisine but also marries western cuisine as well.

主料 Main materials

羊肩肉，羊鞍肉。

lamb shoulder, lamb saddle.

配料 Ingredients

菠菜，青豆，胡萝卜，芹菜根，香芹末，葱段，姜片，干番茄，蒜末，土豆，藜麦蛋挞壳。

spinach, green peas, carrots, celeriac, minced parsley, green onion section, sliced ginger, dried tomatoes, minced garlic, potatoes, quinoa tart shells.

调料 Seasonings

黄油，色拉油，橄榄油，意大利黑醋，奶油，红糖，黑胡椒，酱油，黑蒜粉，葱姜汁，鸡粉，盐，粘肉粉，鸡汤，淀粉，茴香，大料，花椒，桂皮，香叶，百里香，迷迭香，酱汁。

butter, salad oil, olive oil, balsamic vinegar, cream, brown sugar, black pepper, soy sauce, black garlic powder, onion and ginger juice, chicken powder, salt, meat glue powder, chicken broth, starch, fennel, star anise, peppercorns, cinnamon, bay leaves, thyme, rosemary, sauce.

制作过程 Methods

❶ 将羊鞍肉去除边角料，切成两个长条，其中一条裹匀黑蒜粉和粘肉粉，另一条撒上黑胡椒、盐备用。将取下的边角料剁成肉馅，加入葱姜汁、酱油、盐、鸡粉、淀粉调成馅料，均匀平铺在保鲜膜上，再将两个羊肉条并排摆在上面，然后卷起保鲜膜，卷成圆柱形，装入袋子然后真空包装，

放入冷藏室冷却30分钟以上，取出放入50℃的恒温水槽中50分钟。

Remove trimmings from the lamb saddle and cut into two long strips, one of which is coated with black garlic powder and meat glue powder, and the other one is sprinkled with black pepper and salt for later use. Grind the trim to the filling, add onion and ginger juice, soy sauce, salt, chicken powder and starch. Spread them evenly on the saran wrap, put two pieces of lamb saddle side by side on top, and roll them up with saran wrap into a cylinder shape. Put it into a bag and then vacuum seal it. Put it in the refrigerator for at least 30 minutes, take it out and put it into the sous vide at 50℃ for 50 minutes.

❷ 锅中加入水，放入羊肩肉、葱段、姜片、茴香、大料、花椒、桂皮、香叶、酱油、红糖、盐、鸡粉，大火烧开，改成小火煮至羊肩肉熟透，筷子可以穿透羊肩肉即可。取出羊肩肉撕成小块，摆在方形的不锈钢盘子中，上面用重物压实，放入冷藏室冷却1小时，取出切成长方块。取原汤用淀粉勾芡，浇在羊肩肉上，放入70℃的烤箱重新加热即可。

Add water to the pot, add lamb shoulder, green onion section, sliced ginger, fennel, star anise, pepercorns, cinnamon, bay leaves, soy sauce, brown sugar, salt and chicken powder, bring to a boil over high heat, then turn to low heat and cook until the lamb shoulder is cooked and tender. To test it, a pair of chopsticks should be able to penetrate the lamb shoulder. Pull out the lamb shoulder and tear it into small pieces, place them on a square stainless steel plate mold, press it with a heavy weight, then put it in the fridge to cool for 1 hour. Take it out and cut it into long squares. Take the original soup and thicken it with starch, pour it on the lamb shoulder, and reheat it in the oven at 70℃.

❸ 将土豆去皮洗净，用模具切成圆柱状。锅内加入黄油、鸡汤、盐、黑胡椒和土豆，小火煮至土豆熟透取出。平底锅倒入色拉油加热，放入蒜末、百里香、迷迭香，再摆入土豆，将上下两端煎至金黄色备用。

Peel and wash the potatoes and cut them into cylinders with a mold. Put butter, chicken broth, salt, black pepper and potatoes into the pot, cook on low heat until the potatoes are cooked. Heat salad oil in a pan, add minced garlic, thyme, rosemary and potatoes, sear the top and bottom until golden brown color.

4. 将胡萝卜和芹菜根去皮洗净，切成丁状。将青豆和菠菜焯水。干番茄切成小粒备用。

   Peel and wash the carrots and celeriac and cut into cubes. Blanch the green peas and spinach. Cut the dried tomatoes into small pieces for later use.

5. 锅中加入黄油、奶油、盐、胡萝卜、芹菜根煮透捞出，撒入香芹末。锅中加入黄油，加入蒜末煸香，放入一部分青豆、菠菜翻炒，加入鸡汤，加盐调味，倒入搅拌机内搅拌成泥状，取出备用。

   Add butter, cream, salt, carrots and celeriac to the pan, cook thoroughly, remove and sprinkle with minced parsley. Add butter to the pot, swear minced garlic, add some green peas, spinach and chicken broth. Season with salt and pour into a blender and blend into a puree. Take it out and set aside.

6. 切好的番茄粒和另一部分青豆加入橄榄油、盐、意大利黑醋、黑胡椒拌匀，装进藜麦蛋挞壳内，挤入青豆菠菜泥。

   Mix the chopped tomatoes and another part of green peas with olive oil, salt, balsamic vinegar and black pepper. Put them into the quinoa tart shell, squeeze green peas spinach puree.

7. 将以上的成品摆盘，先放青豆菠菜泥，再依次放藜麦蛋挞壳、方旦土豆、酱羊肩肉、黑蒜羊鞍肉卷、胡萝卜和芹菜根丁，最后放上酱汁。

   Place the green peas spinach puree first, followed by the quinoa tart shell, fondant potatoes, lamb shoulder with sauce, black garlic lamb saddle roll, carrots and celeriac. Lastly, add the sauce.

技术解析 Technical resolution

1. 羊鞍肉放入粘肉粉之后，一定要放入冷藏室中冷却30分钟以上，这样粘肉粉才能发挥它的作用。

   Lamb saddle must be put in the fridge for at least 30 minutes after adding the meat glue powder. Otherwise, the meat glue powder will not work well.

2. 土豆不宜煮得太久，以牙签可以轻轻穿过为宜。

   Do not let the potatoes cook for too long. To test them, a toothpick should be able to easily pass through them.

3. 保存青豆菠菜泥的温度不要超过60℃，以免变色。

   The temperature of green peas spinach puree should not exceed 60°C to avoid discoloration.

酱汁制作
Making sauces

锅中加入鸡汤，用微火熬至汤汁黏稠，加入黄油、盐、黑胡椒粉调味即成酱汁。

Put chicken broth in the pot and simmer until reduced to desired consistency. Add butter, salt and black pepper powder to taste.

四、煨海参煎饺佐食青豆泥汁
No.4 Braised Sea Cucumber and Pan-fried Dumpling with Green Peas Puree

小记 Notes

此菜制作的理念来自中餐的烩海参和煎海参饺。中式烹调的理念中没有办法将烩海参和煎海参饺放在一碟菜中呈现给顾客。而西餐却恰恰相反，它可以用菜品和面食混搭或者加以其他成分，制作成美味的前菜。煨海参煎饺佐食青豆泥汁就是中餐的烹调方法结合西餐的理念制作的一道美味的西餐前菜。

This dish is inspired by braised sea cucumber and pan-fried sea cucumber dumplings in Chinese cuisine. The idea of Chinese cooking cannot present braised sea cucumber and pan-fried sea cucumber dumplings to customers in one dish. On the contrary, western cuisine can be made by mixing dishes with pasta or adding other ingredients to create a delicious appetizer. Braised sea cucumber and pan-fried dumpling with green peas puree is a delicious western appetizer made by combining the cooking method of Chinese cuisine with the idea of western cuisine.

主料 Main materials

即食海参，猪肉馅。

instant sea cucumber, pork filling.

配料 Ingredients

青豆，秋葵，葱段，姜片，蒜片，葱姜末，香菜末，罗勒叶，饺子皮，洋葱末，蒜末。

green peas, okra, green onion section, sliced ginger, sliced garlic, minced green onion and ginger, minced cilantro, basil leaves, dumpling wrappers, minced onion, minced garlic.

调料 Seasonings

黄油，芥花油，香油，鲍鱼汁，味极鲜酱油，鸡粉，盐，糖，高汤，湿淀粉，奶油，鸡汤，芥末苏马克粉。

butter, canola oil, sesame oil, abalone sauce, Weijixian soy sauce, chicken powder, salt, sugar, stock, wet starch, cream, chicken broth, mustard sumac powder.

制作过程 Methods

❶ 锅中倒入芥花油烧热，放入葱段、姜片、蒜片煸香，加入高汤、鲍鱼汁、味极鲜酱油、盐、糖、鸡粉，烧沸后放入一些海参转小火煨至入味，用湿淀粉勾芡。

Heat the canola oil in a pan, add green onion section, sliced ginger, sliced garlic and sweat until soft. Add stock, abalone sauce, Weijixian soy sauce, salt, sugar and chicken powder. Bring to a boil, then add parts of sea cucumbers, turn to low heat and simmer until tasty. Thicken with wet starch.

❷ 将剩余没有煨制的海参切成粒状，与猪肉馅、味极鲜酱油、盐、鸡粉、葱姜末、香菜末、香油调匀成馅料，包成饺子。平底锅内放入芥花油，放入饺子，加入少许水，小火煎至汤汁干掉，饺子底部成金黄色即可关火。

Finely dice the remaining sea cucumbers and mix them with pork filling, Weijixian soy sauce, salt, chicken powder, minced green onion and ginger, minced cilantro and sesame oil to make fillings for the dumplings. Place the canola oil in a frying pan, add sea cucumber dumplings and a little water over low heat until the water completely evaporates. Turn off the heat when the bottom of the dumplings turns golden brown.

❸ 锅中加入黄油、洋葱末、蒜末煸香，加入青豆、罗勒叶、奶油、鸡汤、盐调味，略煮，倒入搅拌机内搅拌成泥状，取出保存在60℃的盛器内。

Add butter, minced onion and minced garlic to the pot and sweat until soft. Add green peas, basil leaves, cream, chicken broth and salt to taste. Cook for a while, pour it into a blender and blend into a puree, take it out and store it in a container at 60°C.

❹ 将秋葵一剖为二，焯盐水，捞出在刀口面均匀撒上芥末苏马克粉。

Cut okra into two pieces, blanch in salt water, remove and sprinkle with mustard sumac powder evenly on the cutting surface.

❺ 将以上的成品按照图片装盘即可。

Follow the picture above to plate the dish.

技术解析 Technical resolution

保存青豆泥汁的温度不要超过60℃，以免变色。

The temperature of the green peas puree should not exceed more than 60°C to avoid discoloration.

五、脆香开心果烤海参佐食意大利炖饭
No.5 Crispy Pistachio Roasted Sea Cucumber with Risotto

小记 Notes

此菜制作的理念来自中餐的海参捞饭。用意大利炖饭代替白米饭，再加上用焦脆开心果裹匀的海参，通过中西餐的融合，体现了海参的独特风味。

This dish is inspired by braised sea cucumber with steamed rice in Chinese cuisine. However, the white rice is replaced by risotto and the sea cucumber is wrapped in crispy pistachios. Through the combination of Chinese and western cuisine, this dish embodies a unique take on sea cucumber.

主料 Main materials

泡发海参，黑香米，阿尔博里奥米。

soaked sea cucumber, black rice, Alboriomi.

配料 Ingredients

青豆，青葱末，姜末，蒜末，圆葱末，开心果碎，青豆苗芽。

green peas, minced green onion, minced ginger, minced garlic, minced shallot, chopped pistachios, green pea sprouts.

调料 Seasonings

黄油，芥花油，白葡萄酒，胡椒，帕玛森奶酪，海鲜酱，黑椒汁，花雕酒，鸡粉，盐，高汤。

butter, canola oil, white wine, pepper, parmesan cheese, seafood sauce, black pepper sauce, Huadiao cooking wine, chicken powder, salt, stock.

制作过程 Methods

1. 锅中倒入芥花油烧热，放入青葱末、姜末、蒜末煸香，加入海鲜酱、黑椒汁、花雕酒、鸡粉，小火煮透制作成海参烧烤酱备用。

   Heat the canola oil in a pot, add minced green onion, minced ginger and minced garlic and sweat until soft. Add seafood sauce, black pepper sauce, Huadiao cooking wine and chicken powder, then cook over low heat to make sea cucumber barbecue sauce for later use.

2. 用不锈钢扦子穿好泡发的海参，将海参表面刷匀烧烤酱，用小火烤制2分钟，再刷一层烧烤酱，再烤制3分钟，然后再刷一层烧烤酱，并裹上开心果碎，用小火烘烤，直至开心果焦脆。

   Use a stainless-steel skewer to skewer the soaked sea cucumbers, brush the surface of the sea cucumbers with barbecue sauce, roast on low heat for 2 minutes, brush with a layer of barbecue sauce, and roast for another

3 minutes. Then brush with last layer of barbecue sauce, coated evenly with chopped pistachios, and roasted over low heat until the pistachios are crispy.

3. 黑香米用温水浸泡10分钟捞出备用。锅中加入黄油和圆葱末煸香，加入阿尔博里奥米，小火炒至米粒呈透明，加入黑香米，倒入白葡萄酒炒至酒精挥发，加入青豆和少许高汤，用小火慢煮至米吸收高汤后，再多次重复加入高汤用小火慢煮直到米饭熟透关火，加入黄油、磨碎的帕玛森奶酪、盐、胡椒碎调味。

   Soak black rice in warm water for 10 minutes, remove and set aside. Add butter to the pan, then add minced shallot and sweat until they are fragrant. Add Alboriomi into the pan and saute on low heat until the rice grains are transparent. Add black rice, white wine and continue to saute until the alcohol evaporates. Add green peas and a little stock and cook on low heat until the rice absorbs the stock, then add the remaining stock and cook on low heat until the rice is cooked, then turn off the heat. Add butter, grated parmesan cheese, salt and pepper to taste.

4. 将以上的成品按照图片所示装盘，用青豆苗芽装饰即可。

   Follow the picture above to plate the dish. Garnish with green pea sprouts.

技术解析 Technical resolution

1. 意大利炖饭要使用意大利阿尔博里奥米或者卡纳罗利米。

   The risotto will use either the Italian Alboriomi or Canarolimi.

2. 煮米的时候要注意火候，米煮熟透即可。

   When cooking rice to pay attention to the heat, rice cooked thoroughly.

六、鸡腿肉卷和鸡肉串烧佐食奶酪饺
No.6 Chicken Thigh Roll and Chicken Skewers with Cheese Dumpling

小记 Notes

此菜在2018年迪拜国际美食大赛荣获金牌，融合了中西式的烹饪手法，展示了两种不同的烹调风格。菜品的口味多元化，特别适合北美人的味蕾。

The dish won a gold medal at the 2018 Dubai International Culinary Competition. It is a fusion of Chinese and western cooking techniques, displaying both cooking styles in one dish. Due to the flavorful and complementary nature of this dish, this dish is suitable for a variety of audiences, especially North Americans.

主料 Main materials

带皮无骨鸡腿肉，牛培根。

boneless chicken thigh with skin, beef bacon.

配料 Ingredients

大米，柠檬，白蘑菇，杏子干，红梅干，葡萄干，芦笋，蟹味菇，彩虹甜菜根，紫菜花，彩色饺子皮，香芹末。

rice, lemon, white mushroom, dried apricot, dried prune, raisin, asparagus, hypsizygus marmoreus, rainbow beetroot, purple cauliflower, colorful dumpling wrapper, minced parsley.

调料 Seasonings

欧防风根泥，酱汁，黄油，盐，黑胡椒，鲜奶油，花雕酒，葱姜汁，烧烤酱，瑞可塔奶酪，糖醋汁，圆葱末，芝麻。

parsnip puree, sauce, butter, salt, black pepper, fresh cream, Huadiao cooking wine, onion and ginger juice, barbecue sauce, ricotta cheese, sweet and sour sauce, minced shallot, sesame.

制作过程 Methods

1. 将白蘑菇、水果干剁碎，放入锅内加黄油、盐、黑胡椒、鲜奶油炒香成馅料备用。

   Chop the white mushrooms and dried fruit. Add butter, salt black pepper and fresh cream to stir fry until fragrant. Set aside.

2. 将鸡腿肉去皮，鸡皮放入烤箱烤至酥脆，将部分鸡腿肉用刀片成大片，撒上盐、黑胡椒，放在铺好的牛培根上，再放入炒好的馅料卷起来，用保鲜膜卷紧。放进真空袋里，再放入70℃恒温水槽，煮40分钟，取出用黄油煎至表面金黄。

Separate the chicken skin and meat, put the chicken skin into the oven and roast until crispy. Butterfly the meat into large pieces, sprinkle with salt and black pepper, put it on top of the beef bacon, then put the fried stuffing on top of the chicken and roll it up. Wrap it tightly with saran wrap. Put it in a vacuum bag and seal, then place it into the sous vide and cook at 70℃ for 40 minutes. Take it out and sear with butter until the surface is golden color.

❸ 将剩余的鸡腿肉一部分切成块，用盐、葱姜汁、花雕酒腌制，穿成鸡肉串，在烧烤炉上烤熟，刷一层烧烤酱。

Cut the remaining chicken into small pieces, marinate them with salt, onion and ginger juice and Huadiao cooking wine. Thread the chicken evenly onto the skewers and grill them on the grill. Brush the barbecue sauce.

❹ 另一部分鸡腿肉剁成泥，加入瑞可塔奶酪、盐、柠檬皮末、黑胡椒、香芹末调成饺子馅，放入饺子皮内，包成饺子煮熟捞出，放入融化的黄油内备用。

Grind the remaining chicken into puree, add ricotta cheese, salt, lemon zest, black pepper and minced parsley to make the dumpling filling. Use dumpling wrapper to wrap the filling. Make dumplings and cook them out. Put them in melted butter for later use.

❺ 将芦笋、蟹味菇、紫菜花洗净焯水，锅内放入黄油，加入圆葱末炒香，放入焯水后的蔬菜，加入盐、黑胡椒调味。彩虹甜菜根切薄片加入糖醋汁腌制10分钟。

Wash and blanch asparagus, hypsizygus marmoreus and purple cauliflower. Put butter in the pot and sauté minced shallot until soft. Add vegetables and season with salt and black pepper to taste. Thinly slice the rainbow beetroot in sweet and sour sauce for 10 minutes.

❻ 大米加水、盐煮熟，卷成圆条裹匀芝麻。

Cook rice with water and salt, roll into a round bar and wrap sesame.

❼ 将以上制作的食料摆在盘内，用欧防风根泥和酱汁连接呈现图中的样子即可。

Place the above ingredients on a plate and use the parsnip puree to combine with the sauce as shown in the picture.

技术解析 Technical resolution

❶ 欧防风根泥不宜长时间搅拌。

Parsnip puree does not need to blend for a long time.

❷ 低温慢煮鸡肉，一定要确保鸡肉的内部温度在70℃以上。

Cook the chicken at a low temperature. Make sure the internal temperature of the chicken is above 70°C.

❸ 酱汁要放入冷黄油来增加黏稠度。酱汁黏稠度检测时，取一不锈钢汤匙，将背面蘸上酱汁，酱汁能均匀、薄薄地挂在汤匙即可。

Add cold butter into the sauce to increase the thickness of the sauce. Check the thickness of the sauce by taking a stainless-steel spoon and dipping it into the sauce. The sauce should be able to evenly cover the spoon in a thin coat.

酱汁制作
Making sauces

（1）欧防风根泥制作

Making parsnip puree

欧防风根去皮切成小块，放入锅内加入黄油、鲜奶油、盐、白胡椒、百里香，小火加热至锅内奶油剩下1/3，离火倒入搅拌机内，搅至细腻没有颗粒。

Peel the parsnip and cut into small pieces. Add butter, fresh cream, salt, white pepper and thyme to the pot over low heat to reduce 2/3 of the cream in the pot, then pour it into a blender and blend until a smooth puree is achieved.

（2）酱汁制作

Making sauce

净锅放入褐色鸡汤，小火煨至汤略微黏稠，过滤加入冷黄油搅拌均匀。

Place the brown chicken broth in the saucepan and simmer, and reduce until slightly viscous. Strain and add cold butter while mixing well.

七、脆炸比目鱼柳搭配鱼饺三文鱼沙拉

No.7 Deep Fried Halibut and Fish Dumpling with Salmon Salad

小记 Notes

此菜是一道以西餐味型为主，搭配中餐烹饪技术的中西融合菜品，展现出海鲜的独特风格。

This dish is a fusion of western and Chinese cuisine. By using Chinese cooking techniques while accentuating western flavors, this dish provides a unique take on seafood presentation.

主料 Main materials

比目鱼柳，三文鱼柳。

halibut fillet, salmon fillet.

配料 Ingredients

豌豆粒，干羊肚菌，玉米粒，飞鱼子酱，鱼松。

peas kernels, dried morels, corn kernels, flying fish roe sauce, fish floss.

调料 Seasonings

芥花油，黄油，松露油，罗勒叶，盐，糖，黑胡椒，鲜奶油，美乃滋，蒜末，白胡椒粉，黑胡椒粉，鸡蛋清，面包糠，面粉，澄粉，淀粉，柠檬，香芹，茴香，粘肉粉，酱汁。

canola oil, butter, truffle oil, basil leaves, salt, sugar, black pepper, fresh cream, mayonnaise, minced garlic, white pepper powder, black pepper powder, egg white, breadcrumbs, flour, wheat starch, starch, lemon, parsley, fennel, meat glue powder, sauce.

制作过程 Mcthods

❶ 将比目鱼柳去其边角料，一部分切成长方块，将边角料切成小拇指粗细的长条，均匀撒上盐和黑胡椒粉。将鱼肉裹上面粉、鸡蛋清和面包糠，放入热油锅炸至表面金黄，内部温度为50℃即可。

Trim the halibut fillets and cut them into rectangular shapes. Cut the trimming into long strips with the thickness of a little finger. Sprinkle salt and black pepper evenly. Coat the fish meat with flour, egg white and breadcrumbs. Fry it in hot oil until golden brown, ensuring the internal temperature reaches 50°C.

❷ 将余下的鱼肉放入料理机内，加入鸡蛋清、鲜奶油、柠檬汁、柠檬皮末、白胡椒粉、盐搅拌成鱼胶取出，放入剁碎的茴香末拌匀，放入冷藏室备用。

Put the remaining fish meat into a food processor and add egg white, fresh cream, lemon juice, lemon zest, white pepper powder and salt. Grind until puree. Take it out and mix with chopped fennel and place into the fridge.

❸ 将三文鱼柳切成小拇指粗细的长条，一半用盐、糖、松露油腌制25分钟。用冷水洗去三文鱼表面的盐和糖，吸干水分备用。另一半三文鱼装入真空袋子，放入40℃恒温水槽中45分钟，取出放入冰水中过凉，切成小粒备用。

Cut the salmon into long strips with the thickness of a little finger. Marinate half the portion with salt, sugar and truffle oil for 25 minutes. Wash away the salt and sugar on the surface of the salmon with cold water and dry it. Put the other half of the salmon in a vacuum bag and put it in a 40°C water tank for 45 minutes. Take it out and put it in ice water to cool down. Finely dice it for later use.

❹ 将比目鱼条和三文鱼条放入托盘内，均匀撒上一层薄薄的粘肉粉，然后将比目鱼条和三文鱼条交错地叠在一起，放到长方形模具中盖住盖子，真空包装。随后放入50℃恒温水槽中45分钟，取出再放入急冻冰箱冷冻1小时。

Put halibut strips and salmon strips into a tray and sprinkle a thin layer of meat glue powder evenly. Then stack halibut strips and salmon strips alternately together and put them in a rectangular mold. Cover with a lid and vacuum package. Then put it in a 50°C water tank for 45 minutes. Take it out and put it in a blast chiller for 1 hour.

❺ 碗中放入澄粉、淀粉，一边搅拌一边加入热水，直到没有干面粉，盖上盖子闷几分钟，然后揉匀，加入芥花油继续揉匀，揉匀后静置15～30分钟，然后用刀切成几份，做成饺子皮，包裹先前做好的鱼胶，制成鱼饺，放入蒸箱蒸熟，取出抹上黄油，撒上香芹末。

Put wheat starch and starch in a bowl and stir while adding hot water until there is no dry flour left. Cover it for a few minutes, then knead evenly and add canola oil to continue kneading evenly. Let the dough rest for

15 to 30 minutes. Cut it into portions and press it down using a knife to make thin dumpling wrappers. Then wrap the previously made fish puree to make fish dumplings. Steam until cooked. Take it out and spread butter on top and sprinkle with minced parsley.

6. 将玉米粒、豌豆粒分别焯水过凉，干羊肚菌泡水。

   Blanch corn kernels and peas kernels separately in boiling water and cool them down. Soak dried morels in water.

7. 锅内放入黄油、蒜末炒香，放入羊肚菌、鲜奶油略烧，再放入豌豆粒，用盐、黑胡椒调味。

   Put butter in a pan and sauté minced garlic until soft. Add morels and fresh cream to cook slightly before adding peas for seasoning with salt and black pepper.

8. 将玉米粒和切好的三文鱼粒放在不锈钢碗内，用柠檬汁、美乃滋、盐、切丝的罗勒叶拌匀成三文鱼沙拉。取出急冻冰箱的鱼条，然后用刨片机切成薄片卷成圆柱形，圆柱中间装满三文鱼沙拉，再放上飞鱼子酱、鱼松、罗勒叶。

   Put corn kernels and small diced cooked salmon in a stainless-steel bowl. Mix them well with lemon juice, mayonnaise, salt and shredded basil leaves to make salmon salad. Take out the frozen pressed fish strips. Use the slicer to cut them into thin slices and roll them into cylinder shapes. Fill the cylinder with salmon salad and add flying fish roe sauce, fish floss and basil leaves.

9. 将以上成品依次装盘，浇酱汁即可。

   Put the finished products on a plate and pour over the sauce.

技术解析 Technical resolution

1. 鱼肉加工时要严格按照规定的温度进行恒温和冷冻处理。

   Fish processing in strict accordance with the provisions of the temperature for constant temperature and freezing treatment.

2. 制作鱼饺的面皮要充分静置揉匀。

   The dough for making fish dumplings should be kneaded well.

酱汁制作
Making sauces

净锅放入鲜奶油、咖喱块小火慢煮，待咖喱全部融化时，用盐调味。

Put fresh cream and curry cube into a clean pot over low heat to cook slowly until all the curry melts, then season with salt.

八、低温慢煮猪里脊卷榨菜香菇配酱香黄豆

No.8 Sous Vide Pork Tenderloin with Miso Soybeans

小记 Notes

加拿大猪肉本身的腥味很重，猪肉在西餐中相对其他肉类使用得较少一些，因此此菜采用酱香排骨的概念，融合了西式的低温慢煮烹饪手法去其猪腥味，再搭配中式配料佐食。

Canadian pork has a strong smell, so it is used less in western cuisine compared to other meats. Therefore, this dish adopts the concept of sauce-flavored pork ribs, which incorporates western sous vide techniques to remove its smell. Additionally, Chinese ingredients are added to create a unique fusion dish.

主料 Main materials

猪里脊。

pork tenderloin.

配料 Ingredients

圆葱，榨菜，香菇，胡萝卜，黄豆，土豆，香芹，青葱粒，淀粉。

shallot, pickled mustard root, shiitake mushrooms, carrots, soybeans, potatoes, parsley, chopped green onion, starch.

调料 Seasonings

黄油，芥花油，奶油，盐，糖，大料，葱，姜，蒜，花椒，黑胡椒粉，茴香，花雕酒，白酱油，鸡粉，食用粘合粉，甜面酱，红梅酱，黄芥末酱，酱汁，猪骨汤。

butter, canola oil, cream, salt, sugar, star anise, onion, ginger, garlic, pepper, black pepper powder, fennel, Huadiao cooking wine, white soy sauce, chicken powder, edible binding powder, sweet soya paste, red plum jam, yellow mustard sauce, sauce, pork broth.

制作过程 Methods

❶ 锅中放入猪骨汤、葱、姜、大料、花椒、茴香、花雕酒、白酱油、鸡粉，烧开凉凉。倒进密封袋中。猪里脊洗净，放到盛有调好猪骨汤的密封袋中腌制3小时。

Place pork broth, onion, ginger, star anise, pepper, fennel, Huadiao cooking wine, white soy sauce and chicken powder into the pot, boil, then let it cool. Wash the pork tenderloin and marinate it in a sealed bag filled with brine for 3 hours.

❷ 榨菜切粒，香菇切粒，放入锅内加黄油、盐、黑胡椒粉炒香，放到平铺的猪里脊肉上，卷起来呈

圆形，用绳子捆紧，放入真空袋内，放入70℃的恒温水槽中煮12小时取出。平底锅内加入黄油，取出猪里脊煎至表面金黄，刷上一层薄薄的黄芥末酱，撒上青葱粒。

Finely dice the pickled mustard root and shiitake mushrooms. Melt butter in the pan, add salt and black pepper powder to taste. Flat the pork tenderloin, put the sautéed vegetables on top of the pork and roll it up, tie the twine tightly, put it in a vacuum bag and seal. Place in the water tank and cook at 70°C for 12 hours. Remove the bag and pat dry the meat. Melt butter in the pan, place the pork tenderloin and sear until the surface is golden color. Brush with a thin layer of yellow mustard sauce and sprinkle with chopped green onion.

❸ 黄豆放入冷水中涨发，锅内入芥花油煸香葱、姜、甜面酱、鸡粉、猪骨汤、黄豆煨至入味，加入淀粉勾芡。撒上香芹末。

Soak soybeans in cold water overnight. Put canola oil in the pan and sauté the onion, ginger, sweet soy paste, chicken powder, pork broth, soybeans, and simmer until tasty. Add starch to thicken it. Sprinkle with minced parsley.

❹ 将胡萝卜去皮。锅内加入黄油，加入蒜末，煸香，加入水和胡萝卜，水刚刚淹过胡萝卜，加盐、黑胡椒粉调味，用中火烧至水干，放入食用粘合粉拌匀，胡萝卜交叉排列，整齐地摆放在长方形模具中，顶部用重物压实。放入冰箱冷藏2小时，取出切成长方形条，放入50℃的烤箱内加热5分钟。

Peel the carrots. Melt butter in the pan and sweat minced garlic until soft. Add carrots and water. Add just enough water to cover the carrots. Add salt and black pepper powder to it over medium heat until the water evaporates. Add the edible binding powder and mix well. Place in a rectangular mold and add a weight to the top so it remains compacted. Place in the fridge for 2 hours, take it out and cut into a rectangular bar. Reheat in the oven at 50°C for 5 minutes.

❺ 将土豆去皮切薄片，加入盐、奶油、黄油、香芹末搅拌均匀，然后一层层叠起来，盖上锡纸放入烤箱烤至熟透，去掉锡纸继续烤至土豆金黄，取出在土豆表面放上红梅酱。

Peel and cut the potatoes into thin slices. Add salt, cream, butter, minced parsley and toss evenly, then stack them up layer by layer. Cover with the foil and bake in the oven until they are fully cooked. Remove the foil and continue to bake the potatoes until they are golden color. Then take them out and put red plum jam on the surface of the potatoes.

❻ 将圆葱去皮，放入锅中煎至两边略焦，然后加盐、糖、水、黄油煮至圆葱表面裹一层汁液。

Peel the shallot and fry them in a pan until they are slightly tender on both sides, then add salt, sugar, water, butter and cook until the shallot is covered with a layer of juice.

❼ 将以上的成品按图所示装盘，最后放上酱汁。

Follow the picture above to plate the dish. Lastly, add the sauce.

技术解析 Technical resolution

❶ 酱汁中红葡萄酒的酒精一定要完全挥发掉。

The alcohol in the red wine must be completely evaporated to make the sauce.

❷ 低温慢煮猪肉时，一定要确保猪肉内部的温度达到70℃。

When low temperature slow cooking the pork, ensure its internal temperature reaches 70°C.

酱汁制作
Making sauces

净锅放入红葡萄酒，加热至酒精挥发，加入煮猪里脊的猪骨汤大火烧开，小火煨至汤略微黏稠，过滤加入冷黄油搅拌均匀。

Place the red wine in the saucepan, heat until the alcohol evaporates, then add the pork broth that cooked the pork tenderloin. Bring to a boil over high heat, simmer, and reduce until slightly viscous. Strain, then add cold butter while mixing well.

九、阿尔伯塔牛肉双拼佐食芹菜根泥

No.9 Roasted Alberta Beef Tenderlion and Deep-fried Beef Meat Ball with Celeriac Puree

小记 Notes

加拿大阿尔伯塔省牛肉是加拿大质量最好的牛肉。此菜选用牛里脊和牛腩两部分，融合西式的烹饪手法和中式的香料及调料创意而成。

Alberta beef is the best quality beef in Canada. This dish uses two parts of beef tenderloin and beef brisket. It is created by combining western cooking techniques with Chinese spices and seasonings.

主料 Main materials

牛里脊，牛腩。

beef tenderloin, beef brisket.

配料 Ingredients

芹菜根，胡萝卜，土豆，香芹末，圆葱末。

celeriac, carrots, potatoes, minced parsley, minced shallot.

调料 Seasonings

芥花油，黄油，盐，白胡椒，奶油，十三香，孜然粉，桂皮，大料，葱，姜，花雕酒，酱油，清牛汤，鸡蛋液，面包糠，胡萝卜汁，奶酪，酱汁。

canola oil, butter, salt, white pepper, cream, mixed spices, cumin powder, cinnamon, star anise, green onion, ginger, Huadiao cooking wine, soy sauce, clear beef broth, egg liquid, breadcrumbs, carrot juice, cheese, sauce.

制作过程 Methods

❶ 将牛里脊撒上十三香、孜然粉、盐、芥花油腌制1小时，然后放入烤箱烤至牛肉内部温度为60℃，取出放置5～10分钟，再切割装盘。

Sprinkle mixed spices, cumin powder, salt and canola oil onto the beef tenderloin and marinate for 1 hour. Then bake in the oven until the internal temperature of the beef is 60℃. Take it out and let it sit for 5 to 10 minutes before cutting it into pieces and plating it.

❷ 将牛腩放入锅中，加入清牛汤、酱油、桂皮、葱、姜、大料、花雕酒、盐，卤至牛腩软糯，取出凉凉，切碎制作成牛肉丸，裹匀蛋液和面包糠炸至金黄色捞出备用。

Put the beef brisket into a pot with clear beef broth, soy sauce, cinnamon, green onion, ginger, star anise, Huadiao cooking wine and salt. Cook until the beef brisket is soft then take it out to cool. Cut it into small pieces to make beef balls. Coat them evenly with egg liquid and breadcrumbs and fry them until golden brown. Set it aside.

❸ 将胡萝卜去皮，放入胡萝卜汁中，加黄油、盐煮至熟透。

Peel the carrots and put them in carrot juice with butter and salt until they are cooked through.

❹ 将土豆去皮切薄片，加入盐、奶油、黄油、香芹末搅拌均匀，然后一层层叠起来，上面铺上一层奶酪，盖上锡纸放入烤箱烤至熟透，去掉锡纸继续烤至奶酪略焦。

Peel the potatoes and slice them thinly. Add salt, cream, butter and minced parsley to them. Then stack them layer by layer with cheese on top. Cover with foil and bake in the oven until cooked through. Remove the foil and continue baking until the cheese is slightly burnt.

❺ 净锅加入黄油、圆葱末煸香，放入切成小块的芹菜根，加入清牛汤、盐、白胡椒、奶油，小火加热至锅内奶油剩下1/3，离火倒入搅拌机内，搅拌至细腻平滑松软状即成。

In a clear pan, add butter, minced shallot saute, add cut celeriac, clear beef broth, salt, white pepper, cream, reduce heat to 1/3 of cream. Remove from heat and pour into blender. Stir until smooth and fluffy.

❻ 装盘点缀：用奶酪、胡萝卜铺底，牛肉放在奶酪上，其他食材点缀，浇酱汁即可。

Plate decoration：use cheese and carrots as a base, place beef on top of the cheese, garnish with other ingredients , pour sauce.

技术解析 Technical resolution

❶ 牛里脊在烤制完之后，一定要取出放置5 ~ 10分钟，目的是让牛肉内汁水充分地被牛肉本身吸收，这样切出来的牛肉汁水不会流失。

After roasting the beef tenderloin, be sure to take it out to rest for 5 to 10 minutes so that the juice inside the beef can be fully absorbed by the beef itself. This way the juice won't run out when you cut it.

❷ 芹菜根泥不宜长时间搅拌，否则会比较黏稠。

Celeriac puree should not be blended for too long otherwise it will become too thick.

牛里脊效果图

Rendering of beef tenderlion

酱汁制作
Making sauces

净锅倒入红葡萄酒，煨至一半时加入牛腩卤汁小火煨至汤略微黏稠，过滤加入少量冷黄油混合备用。

Pour red wine into a clean saucepan and simmer to reduce until half. Add some beef brisket broth and simmer over low heat until the soup is slightly thickened. Strain and mix it with a small amount of cold butter.

十、清汤奶酪虾饺
No.10 Shrimp Dumpling Consommé

小记 Notes

此菜荣获2016年德国IKA奥林匹克烹饪大赛国家冷台展铜牌。清汤奶酪虾饺属于此书中馄饨汤的高级版本，制作此汤的理念来自馄饨汤。汤底的制作完全采用西式清汤的做法。同时在传统虾饺中加入少许奶油和奶酪，使虾饺与清汤完美结合。

This dish won the bronze medal of the National Cold Table exhibition at the IKA Olympic Cooking Competition in Germany in 2016. Shrimp dumplings consommé is an advanced version of wonton soup, which is what this dish is originally inspired by. The soup base is completely based on the method of western consommé. At the same time, a little cream and cheese is added to the traditional shrimp dumplings, which perfectly combines with the consommé.

主料
Main materials

虾仁。
shrimp.

配料
Ingredients

油菜心，木耳丝，西蓝花，西芹，圆葱，胡萝卜。
rape heart, shredded fungus, broccoli, celery, shallot, carrots.

调料 Seasonings

橄榄油，鸡汤，丁香，香叶，盐，柠檬汁，蛋清，澄粉，淀粉，葱姜汁，鸡粉，奶油，奶酪。

olive oil, chicken broth, cloves, bay leaves, salt, lemon juice, egg white, wheat starch, starch, onion and ginger juice, chicken powder, cream, cheese.

制作过程 Methods

1. 将虾仁去掉沙线洗净备用。取一半的虾仁，加入葱姜汁、盐、鸡粉、奶油、奶酪，放入搅拌机搅拌成虾蓉备用。余下的虾仁撒少许盐腌制。

   Remove the sand vein off the shrimp and wash it for later use. Take half of the shrimp and add onion and ginger juice, salt, chicken powder, cream and cheese to the blender to make shrimp paste for later use. The remaining shrimp is sprinkled with a little bit of salt.

2. 碗中放入澄粉、淀粉、盐一边搅拌一边加入热水，直到没有干面粉，盖上盖子闷几分钟，然后揉匀，加入橄榄油继续揉匀，取适量澄面擀成圆形的皮，放入虾蓉和一个虾仁，包成饺子形状，放入蒸笼蒸熟。

   In a bowl, stir in the wheat starch, starch and salt and add the hot water until there is no dry flour. Cover and simmer for a few minutes, then knead well. Add the olive oil and continue kneading well. Take some of the flour and roll it into a round skin, put in shrimp paste and a shrimp, wrap into the shape of dumplings, steam in a steamer.

3. 将油菜心和木耳焯水，将西芹、西蓝花、圆葱、胡萝卜切成小粒，放入搅拌好的蛋清中，再放入丁香、香叶、柠檬汁。

   Blanch rape heart and fungus, cut celery, broccoli, shallot carrots into small particles, into the egg white, then add cloves, bay leaves, lemon juice.

4. 将鸡汤加热，倒入蛋清混合物中，烧开，同时不停地搅动，直到蛋清凝固浮起，停止搅动，在凝固的蛋清中戳一小孔，小火炖1小时以上，过滤成清汤。

   Heat the chicken broth and pour it into the egg white mixture and bring to a boil. Stir constantly until the egg white solidifies and floats up. Stop stirring and make a small hole in the solidified egg white. Simmer over low heat for more than 1 hour until it becomes clear.

5. 汤碗中先放入虾饺，再放入油菜心、木耳，最后倒入清汤。

   First, put the shrimp dumpling in the soup bowl, then put in the rape heart and fungus. Lastly, pour in the clear soup.

技术解析 Technical resolution

吊制清汤时要注意，当蛋清凝固时不可以再搅动，如果搅动蛋清使其破碎就达不到预期的效果，西式吊清汤的原理和中餐吊清汤的原理相同，只是底汤的口味不同。

When making consommé, be careful not to stir while the egg white solidifies. If so, it will break up the egg white and not achieve the expected result. The principle of western-style clear soup is the same as that of Chinese-style clear soup. Only the taste of the base soup is different.

十一、葱香原汁海参熏鳕鱼

No.11 Braised Sea Cucumber with Natural Reduction and Smoked Cod

小记 Notes

此菜是在葱烧海参的基础上，加入西式熏鱼制作而成，使菜品呈现典型中西方烹饪技艺的合璧，是中西味蕾的匹配。

This dish is originally inspired by onion-braised sea cucumber and is created by smoking the fish in a western fashion. By combining Chinese and western cooking, this dish will appeal to both cultural taste buds.

主料 Main materials

涨发海参，鳕鱼。

soaked sea cucumber, cod.

配料 Ingredients

菠菜，番茄，香菜叶。

spinach, tomato, cilantro leaves.

调料 Seasonings

芥花油，葱油，黄油，橄榄油，柠檬，枫叶糖浆，松木屑，大料，葱段，生抽，老抽，蚝油，糖，鸡汤，盐，白胡椒粒，白醋，罗勒叶，芝麻酱，鸡精，料酒。

canola oil, onion flavor oil, butter, olive oil, lemon, maple syrup, pine wood chips, star anise, green onion section, light soy sauce, dark soy sauce, oyster sauce, sugar, chicken broth, salt, white pepper, white vinegar, basil leaves, sesame paste, chicken essence, cooking wine.

制作过程 Methods

1. 海参制作过程：将海参焯水。锅内放葱油，烧热后加入葱段、大料爆香，再加入鸡精、料酒、蚝油、生抽、老抽、糖、鸡汤、盐，放入海参，然后盖上锅盖焖至汤汁黏稠，取出海参，淋入少许葱油，过滤取汤汁待用。

   Sea cucumber production process: Blanch the sea cucumber. Put onion flavor oil in the pot and heat it up. Add green onion section and star anise to stir-fry until fragrant. Then add chicken essence, cooking wine, oyster sauce, light soy sauce, dark soy sauce, sugar, chicken broth and salt. Put in the sea cucumber and then cover the pot to simmer until the soup is thick. Take out the sea cucumber and then add some onion flavor oil, then strain the soup for later use.

2. 鳕鱼制作过程：将鳕鱼去鳞，清洗干净，切成1.5厘米厚的鱼排。柠檬切片与盐、白胡椒粒、罗勒叶、芥花油混合，放入鱼排腌制1小时，熏炉内放入松木屑再将其熏25分钟，取出鱼排，用刷子将枫叶糖浆均匀地涂在鱼排上，再放入200℃烤箱烤3分钟，至鱼排表面为金黄色。

Cod production process: Remove the scales from the cod and wash it clean. Cut it into 1.5 cm thick slices. Mix lemon slices with salt, white pepper, basil leaves and canola oil. Put the cod in the marinade for 1 hour. Put it in a pine wood chip smoker for 25 minutes. Take out the cod and brush maple syrup evenly on them. Then put them in a 200℃ oven for 3 minutes until the surface of the cod is golden brown.

❸ 芝麻菠菜制作过程：将菠菜洗净焯水备用，芝麻酱用开水调稀，加入盐、少许生抽、黄油调匀，拌入菠菜即可。

Sesame spinach production process: Blanch spinach in water and set aside. Dilute sesame paste with boiling water and add salt, a little light soy sauce and butter. Mix the sesame sauce with spinach together.

❹ 番茄制作过程：将番茄焯水去皮，切成片加入白醋、盐和橄榄油调匀。

Tomato production process: Blanch tomato to remove skin and cut into slices. Add white vinegar, salt and olive oil.

❺ 将以上所有的原料装盘，用香菜叶点缀即可。

Put all the above ingredients on a platter and garnish with cilantro leaves.

技术解析 Technical resolution

❶ 松木屑每次要少量放置，烟气没有时要尝试味道，然后决定是否继续放松木屑，以免烟熏味过重。

Use pine wood chips sparingly. When there is no more smoke, taste it first before adding more pine wood chips to avoid creating an excessive smoky flavor.

❷ 刷枫叶糖浆时，要均匀的薄薄刷一层。

When brushing maple syrup evenly on the cod, make sure it's a thin layer.

十二、酱焖鲈鱼配鱼饺和鸡蛋米卷
No.12 Braised Soya Bean Paste Sea Bass with Sea Bass Dumpling and Egg Roll

小记 Notes

此菜荣获2017年香港国际美食大赛专业组银牌。此菜在鲁菜胶东菜系酱焖鱼的基础上，融入了西式的配料和装盘方式制作而成，既保留了胶东菜系酱焖鱼的家乡味，又挑动了西方人的味蕾和视觉。

This dish won the silver medal in the professional group at the Hong Kong International Culinary Competition in 2017. Based on the concept of Lu/Jiaodong cuisine braised fish, it is created by adding western ingredients and plating methods. It not only retains the hometown flavor of Jiaodong cuisine braised fish, but it also captures the taste buds and visual senses of western audiences.

主料 Main materials

鲈鱼。

sea bass.

配料 Ingredients

蟹味菇，彩虹甜菜根，青葱，韭菜，熟白米饭，煎蛋皮，鸡蛋。

hypsizygus marmoreus, rainbow beetroot, green onion, garlic chives, cooked white rice, egg roll wrapper, eggs.

调料 Seasonings

黄油，芥花油，葱，姜，蒜，干辣椒，香菜叶，香芹末，香叶，黑胡椒粉，柠檬汁，大料，面酱，酱油，香油，面粉，胡萝卜汁，盐，香炸脆粉，淡奶油，奶酪，高汤。

butter, canola oil, onion, ginger, garlic, dried chili, cilantro leaves, minced parsley, bay leaves, black pepper powder, lemon juice, star anise, bean paste, soy sauce, sesame oil, flour, carrot juice, salt, crispy powder, whipping cream, cheese, stock.

制作过程 Methods

❶ 制作青葱泥：净锅加水放盐烧开，放入青葱至颜色翠绿，捞出放入冰水中过凉，然后捞出切段放入搅拌机内，加入淡奶油，搅拌至细腻没有颗粒时加入黄油继续搅拌至平滑，倒入净锅内小火加热，加盐调味。

Make green onion puree: Add water and salt to the pot and bring to a boil, then add green onion until the color is emerald green. Remove it and put it in ice water to cool. Remove it then chop it. Place it in a blender with whipping cream. Blend until smooth. Add butter and continue to blend until smooth. Pour into a clean pot and heat over low heat. Add salt to taste.

❷ 制作酱焖鱼：将鲈鱼宰杀去皮，切成长10厘米、宽4厘米的段；净锅放入芥花油加热，放入鱼段煎至两面金黄色，放入葱末、姜末、蒜末、干辣椒、大料、面酱、酱油爆香，加入高汤用大火烧开，转小火煨至汤汁黏稠，关火过滤酱汁备用。

Make braised fish: Fillet the sea bass and remove the skin. Cut it 10 cm long and 4 cm wide. Add canola oil to a pot and heat it up. Sear the fish until both sides are golden brown. Add minced onion, minced ginger, minced garlic, dried chili, star anise, bean paste, soy sauce and stir-fry until fragrant. Add the stock and boil over high heat. Simmer and reduce the soup until it becomes thick and sticky over low heat. Turn off the heat and strain the sauce to keep it on the side.

❸ 制作鱼饺：将面粉、鸡蛋、胡萝卜汁和少许盐混合，和面制作成饺子皮备用。鲈鱼肉切成小粒状，韭菜切成粒状，混合后加入酱油、黑胡椒粉、香油调成馅。净锅烧水加盐烧开，放入鱼饺煮

熟捞出，另起锅加入黄油，放入蒜末炒香，倒入鱼饺翻炒均匀即可。

Make fish dumplings: Mix flour, eggs, carrot juice and a little salt together to make dumpling wrappers for later use. Cut sea bass and garlic chives into small pieces. Mix them with soy sauce, black pepper powder and sesame oil to make fish stuffing. Add salt and water in a pot and bring it to a boil. Put in the fish dumplings and cook it. In another pan, add butter and minced garlic to sweat until soft, then add fish dumplings and toss evenly.

4. 制作鸡蛋米卷：将熟米饭、奶酪混合均匀，用煎蛋皮包住成圆柱状。香炸脆粉加水调成脆皮糊。锅内加入芥花油烧至七成热，放入挂有脆皮糊的鸡蛋卷，炸至金黄色捞出，斜切成小段。

   Make crispy egg roll: Mix cooked rice with cheese evenly, then wrap it up with the egg roll wrapper into a cylinder shape. Mix crispy powder with water to make the batter. Coat the egg roll with crispy batter and deep fry them until golden brown. Remove the egg roll and cut them diagonally into small pieces.

5. 净锅放入黄油加热，放入蒜末炒香，加入蟹味菇、彩虹甜菜根、柠檬汁、盐、黑胡椒粉、香叶调味，撒入香芹末。

   Add butter in a pan and heat it up. Then add minced garlic to sweat until soft, followed by hypsizygus marmoreus, rainbow beetroot, lemon juice, salt, black pepper powder and bay leaves to sauté and season. Sprinkle minced parsley on top.

6. 将以上所有的原料装盘，用香菜叶点缀即可。

   Put all the above ingredients on a platter and garnish with cilantro leaves.

技术解析 Technical resolution

1. 用热水烫过青葱后，一定要用冰水冷却。

   After blanching the green onion in hot water, make sure to cool them down with ice water.

2. 用小火加热青葱泥以防变色。

   Heat the green onion puree over low heat to prevent discoloration.

十三、黄菜花泥配脆皮五花肉

No.13 Yellow Cauliflower Puree with Crispy Pork Belly

小记 Notes

此菜是在东坡肉的基础上，融合了西式的烹饪手法和配料制作而成。使菜品的口味既适合华人，也适合西方人的味蕾。

This dish is based on the concept of Dongpo pork while incorporating western cooking techniques and ingredients. The unique flavors of this dish will appeal to both Chinese and western audiences.

主料 Main materials

带皮五花肉。

pork belly with skin.

配料 Ingredients

黄菜花，豆苗芽。

yellow cauliflower, bean sprouts.

调料 Seasonings

芥花油，黄油，鸭油，白糖，盐，淡奶油，枫叶糖浆，百里香，大料，葱，姜，酱油，白胡椒，老汤。

canola oil, butter, duck oil, white sugar, salt, whipping cream, maple syrup, thyme, star anise, onion, ginger, soy sauce, white pepper, old soup.

制作过程 Methods

❶ 制作黄菜花泥：净锅加入黄菜花、淡奶油、盐、白胡椒、百里香小火加热至锅内奶油剩1/3，离火倒入搅拌机内，搅拌至细腻没有颗粒加入黄油继续搅拌至平滑状即成。

Make yellow cauliflower puree: Put yellow cauliflower, whipping cream, salt, white pepper and thyme in a pot and bring it to a boil. Simmer until 1/3 of the cream remains in the pot. Turn off the heat and pour it into a blender. Blend until smooth. Add butter and continue to blend until smooth.

❷ 制作酱汁：净锅放入少量芥花油加热，放入葱、姜、大料、五花肉、酱油、白糖煸至五花肉上色，加入老汤大火烧开，转小火煨至汤汁略微黏稠，过滤加入少量黄油混合备用。

Make sauce: Put a small amount of canola oil into the net pot heat, add onion, ginger, star anise, pork belly, soy sauce, white sugar stir pork belly color, add old soup, bring to a boil, simmer until the soup slightly thick, filter and mix with a little butter. Set aside.

❸ 制作五花三层肉：将盐和白糖按1∶1的比例混合，然后均匀地涂抹在猪肉上腌制1小时。将鸭油

放入150℃的烤炉中融化。用冷水彻底清洗猪肉表面的盐和白糖，吸干水分，放进融化鸭油中，加入百里香，烤制3小时。

Make pork belly: Mix salt and white sugar in a 1∶1 ratio, then evenly cover the pork belly to cure it for 1 hour. Melt duck oil in an oven heated to 150℃. Rinse off the salt and white sugar on the surface of the pork belly with cold water then pat it dry. Put it into melted duck oil, add thyme, and bake for 3 hours.

❹ 取出猪肉放进冷藏室中冷却2小时。净锅放入枫叶糖浆和盐小火加热15～20分钟，至略微黏稠状。将猪肉切成2.5厘米的方块，放入185℃的油锅中炸至猪皮表面酥脆，捞出控油，放入糖浆中裹匀。装盘，用豆苗芽装饰即成。

Take out the pork belly and put it in the fridge for 2 hours. Place maple syrup and salt in a pot and heat it over low heat for 15 to 20 minutes until slightly thickened. Cut the pork belly into 2.5 cm square shapes, fry them in oil at 185℃ until the skin is crispy, remove them from the oil, drain them, then coat them in syrup.

技术解析 Technical resolution

❶ 酱汁要利用五花肉层次不完整的部分来制作。五花肉的层次分布比较均匀部分作为主料备用。

Use the extra parts of the pork belly to make sauce, while the evenly distributed parts of the pork belly will be used as the main meat.

❷ 黄菜花泥在搅拌过程中一定要加黄油来增加菜花泥滑润的口感。

In the process of blending the yellow cauliflower, butter must be added to increase its smooth taste.

❸ 熬制枫叶糖浆要用小火，注意观察，以免火过大，使糖浆颜色变得灰暗。

Carefully boil the maple syrup over low heat and observe carefully to avoid turning the syrup into a dull color.

十四、中式炸鸡配百里香脱脂乳华夫饼
No.14 Chinese-style Fried Chicken with Buttermilk Thyme Waffles

小记 Notes

此菜是将中式的炸鸡块与西式的脱脂乳华夫饼结合制作而成的一道前菜。

This dish is a creative appetizer that combines Chinese-style fried chicken nuggets with western-style buttermilk waffles.

主料 Main materials

无皮鸡腿肉。

skinless chicken thigh meat.

配料 Ingredients

面粉，泡打粉，鸡蛋，玉米淀粉，豆苗芽。

flour, baking powder, eggs, corn starch, bean sprouts.

调料 Seasonings

黄油，白糖，盐，脱脂乳，枫叶糖浆，蜂蜜，干百里香末，葱，姜，花雕酒。

butter, white sugar, salt, buttermilk, maple syrup, honey, dried thyme powder, onion, ginger, Huadiao cooking wine.

制作过程 Methods

1. 制作华夫饼：将面粉、盐、泡打粉、白糖、鸡蛋、脱脂乳、干百里香末混合均匀调制成华夫饼面浆备用，将华夫饼煎锅加热，然后倒入华夫饼面浆，煎至两面金黄，起锅切成小块备用。

   Make waffles: Mix flour, salt, baking powder, white sugar, eggs, buttermilk and dried thyme powder evenly to make the waffle batter. Heat the waffle pan and pour in the waffle batter. Fry until both sides are golden brown. Cut into small pieces and set aside.

2. 制作炸鸡块：将无皮鸡腿肉切成小块，加入盐、葱、姜、花雕酒腌制。加入面粉和玉米淀粉混匀，放入七成热油温炸至外焦里嫩。

   Make fried chicken nuggets: Cut skinless chicken thigh meat into small pieces and marinate them with salt, onion, ginger and Huadiao cooking wine. Add flour and corn starch and mix well. Fry in oil until they are crispy on the outside and tender on the inside.

3. 制作酱汁：净锅放入枫叶糖浆、蜂蜜、黄油低温加热，直到黄油融化，搅拌均匀即可。

   Make sauce:Put maple syrup, honey and butter in a pan and heat it up at low heat until the butter melts and mixes well.

④ 将所有成品装盘，用豆苗芽点缀。

Put all the finished products on a plate, decorate with bean sprouts.

技术解析 Technical resolution

酱汁不要高温加热，低温至黄油融化，搅拌均匀即可。长时间加热会导致酱汁呈浑浊状。

Do not heat the sauce at high temperature. Heat it at low temperature until the butter melts and stirs well. Long-time heating will cause the sauce to become cloudy.

十五、大西洋三文鱼配水萝卜酸甜橙汁
No.15 Atlantic Salmon and Red Radish with Orange Sweet Sour Sauce

小记 Notes

大西洋三文鱼在加拿大被广泛食用，此菜最初的创意来自中餐的五香熏鱼，由于要保持三文鱼的自身颜色和肉质的鲜美，不宜采用高温长时间烹制的方法，所以采用低温慢煮的烹调方法来保持三文鱼肉的本色和鲜嫩。酱汁由中餐的褐色酱汁，改成甜酸橙汁酱，配以清口水萝卜食用。菜品成色感官上符合西餐中的前菜，口味也比较适合北美人，成为客人比较喜爱的冷开胃菜之一。

Atlantic salmon is widely consumed in Canada. This dish was originally inspired by Chinese-style five spices smoked fish. To maintain the natural color and freshness of salmon meat, it is not easy to use high-temperature and long-time cooking methods. Therefore, sous vide is used to control the natural quality and tenderness of salmon meat. The sauce is changed from Chinese-style brown sauce to sweet and sour orange juice sauce, served with refreshing red radish. The color and sensory aspects of the dish are consistent with western-style appetizers, and the taste is also suitable for North Americans, becoming one of the cold appetizers that guests prefer.

主料 Main materials

大西洋三文鱼。

Atlantic salmon.

配料 Ingredients

水萝卜，萝卜苗。

red radish, radish sprouts.

调料 Seasonings

橄榄油，韭葱，五香粉，盐，花雕酒，浓缩橙汁，糖，蜂蜜，白醋。

olive oil, leek, five spice powder, salt, Huadiao cooking wine, concentrated orange juice, sugar, honey, white vinegar.

制作过程 Methods

1. 将韭葱放入烤箱内烤至黑色，用搅拌机搅拌成细粉状备用。
   Roast leek in the oven until black, then grind them into fine powder with a blender and set aside.
2. 将三文鱼宰杀去皮洗净，切成长条，用盐、糖、五香粉、花雕酒腌制1小时。然后用冷水洗去鱼表面的盐、糖、五香粉，吸干水分，将鱼条顺切成三部分，在刀切面均匀撒上五香粉和烘烤的韭葱粉，将鱼条恢复原样，用保鲜膜卷紧成圆形，真空包装，放入41℃的低温慢煮机中煮1小时，捞出连真空袋一起放入冰水中过凉，直到内部温度降至0℃。

Fillet the salmon, remove the skin and wash it clean. Cut it into long strips and marinate it with salt, sugar, five spice powder and Huadiao cooking wine for 1 hour. Then rinse off the salt, sugar and five spice powder on the surface of the fish with cold water and pat dry. Cut the fish strips into 3 parts, sprinkle evenly with five spice powder and leek powder on the cutting edge of the fish strips, then place them back into their original shape. Wrap them tightly with saran wrap, vacuum seal them, then put them in a 41℃ water tank for 1 hour. Take out the fish from the water tank and put them in ice water to cool until the internal temperature is 0℃.

3. 将水萝卜洗净切成薄片，萝卜苗洗净。

   Wash red radish and cut into thin slices. Wash radish sprouts clean.

4. 制作酸酱汁：碗中放入浓缩橙汁、蜂蜜、白醋、橄榄油调和均匀。

   Make sorrel: Put concentrated orange juice, honey, white vinegar and olive oil in a bowl and mix well.

5. 按照图片所示来装盘。用酸酱汁装饰。

   Follow the picture above to plate the dish. Garnish with sorrel.

技术解析 Technical resolution

煮熟的三文鱼冰水过凉时，确保内部的温度在0℃。在切的时候才能保持鱼肉的完整。

When the ice water for cooked salmon is too cold, make sure the internal temperature is 0°C. When cut, the fish will remain intact.

十六、黄油蒜香酸辣龙虾尾五花肉配布里欧面包与荷兰枫糖墨西哥辣椒蛋黄酱

No.16 Butter Garlic Spicy Sour Lobster Tail Pork Belly with Brioche Bread and Dutch Maple Sugar Mexican Chili Mayonnaise

小记 Notes

此菜是在西餐黄油煮龙虾的单一味道中加入中餐的酸辣味，同时用蒜香五花肉取代传统西式培根，混搭中西式的烹饪手法配以西式面包而食的一道前菜。该菜品的口味体现了中餐的烹饪风格，又恰好符合北美人味蕾上对微辣的需求。

This dish combines the sour and spicy flavors of Chinese cuisine with the single taste of western-style butter cooked lobster. At the same time, using garlic-flavored pork belly instead of traditional western-style bacon, this dish is mixed with western and Chinese cooking techniques to create an appetizer that is eaten with western-style bread. The taste of this dish reflects the flavor of Chinese cuisine and also meets the needs of North American's taste buds for a slightly spicy flavor.

主料 Main materials

加拿大龙虾尾，去皮五花肉。

Canadian lobster tail, skinless pork belly.

配料 Ingredients

牛油果，香芹末，食用花瓣，嫩甘蓝叶。

avocado, minced parsley, edible petals, tender cabbage leaves.

调料 Seasonings

黄油，芥花油，橄榄油，味极鲜酱油，龙虾汤，盐，胡椒粒，辣鲜露，墨西哥辣椒酱，枫叶糖浆，蛋黄，橙汁，白醋，柠檬汁，中筋面粉，酵母，糖，鸡蛋，牛奶，大蒜粉。

butter, canola oil, olive oil, Weijixian soy sauce, lobster broth, salt, pepper, spicy liquid seasoning, Mexican chili sauce, maple syrup, egg yolk, orange juice, white vinegar, lemon juice, all-purpose flour, yeast, sugar, eggs, milk, garlic powder.

制作过程 Methods

❶ 将龙虾尾去壳，去沙线。龙虾汤加入盐、胡椒粒、白醋、辣鲜露、黄油加热到53℃保持温度平衡，放入龙虾肉煮35分钟，捞出撒上少许香芹末。

Shell the lobster tail and remove the sand line. Add salt, pepper, white vinegar, spicy liquid seasoning, butter to lobster broth, and heat to 53℃ to balance. Add lobster meat and simmer for 35 minutes. Remove and sprinkle with a little minced parsley.

❷ 用切片机将冷冻的去皮五花肉切成2毫米厚的薄片，用大蒜粉、味极鲜酱油腌制，放入烤箱烤至酥脆。

Use a slicer to slice the frozen skinless pork belly into thin slices 2 mm thick. Marinate with garlic powder and Weijixian soy sauce. Bake in the oven until crispy.

❸ 将牛油果去皮和核，放入搅拌机，加入柠檬汁、橄榄油低速搅拌，使牛油果成为泥状备用。

Peel the avocado and remove the seed. Put it into a blender. Add lemon juice and olive oil and blend at low speed to make the avocado puree for later use.

❹ 制作枫糖墨西哥辣椒蛋黄酱：将蛋黄放入高速功能搅拌机，一边低速搅拌，一边慢慢加入芥花油，当酱汁凝固时加入白醋、枫叶糖浆、墨西哥辣椒酱、盐、橙汁，高速搅拌20～25秒即成。

Make maple sugar Mexican chili mayonnaise: Put egg yolk into a high-speed blender and slowly add canola oil while blending at low speed. When the sauce firms, add white vinegar, maple syrup, Mexican chili sauce, salt, orange juice and blend at high speed for 20 to 25 seconds.

❺ 制作布里欧面包：将面粉、酵母、牛奶、糖、鸡蛋、盐放入和面机内低速搅拌，慢慢加入黄油，中速搅拌均匀，取出放入冷藏室冷却至少18小时，取出所需分量，制作面包坯，刷上蛋液放入200℃烤箱烤12～15分钟。

Make brioche bread: Put flour, yeast, milk, sugar, eggs and salt into the dough mixer and mix at low speed. Slowly add butter and mix evenly at medium speed. Take it out and put it into the fridge for at least 18 hours.

Take out the portion needed and make bread rolls. Brush with egg yolk and bake at a 200°C oven for 12 to 15 minutes.

❻ 装盘，用食用花瓣和嫩甘蓝叶装饰。

Plate and garnish with edible petals and tender cabbage leaves.

技术解析 Technical resolution

❶ 煮龙虾温度要控制在53℃，温度过高，肉质会变硬。

The temperature of poaching lobster should be controlled at 53°C. If the temperature is too high, the meat will become hard.

❷ 牛油果一定要加入柠檬汁，否则会氧化变色。

Avocado must be added with lemon juice; otherwise, it will oxidize and change color.

❸ 枫糖墨西哥辣椒蛋黄酱不宜长时间高速搅拌，高速搅拌会产生热量使酱汁分层，无法融合在一起。

Maple syrup Mexican chili mayonnaise cannot be blended at high speed for a long time. High-speed blending will generate heat and cause the sauce to separate, which cannot be fused together.

十七、炸意大利式松露蘑菇球配糖拌番茄
No.17 Truffle Oil Cheese Mushroom Ball with Sweet Tomato

小记 Notes

此菜是将西式意大利的炸饭团配番茄改成西式炸蘑菇团配中式糖拌番茄，既保留了西式的烹饪手法和配料，又加入中式凉菜制作而成，使菜品成为简单的开胃菜。

This dish is a creative combination of Italian and Chinese cuisine. It takes western fried mushroom balls with Chinese-style sugar-blanched tomatoes to replace the Italian dish of fried rice balls with tomato carpaccio. The dish retains the western cooking techniques and ingredients while adding a simple Chinese-style cold dish, making it a very simple appetizer.

主料 Main materials

蘑菇，番茄。

mushrooms, tomatoes.

配料 Ingredients

圆葱，帕玛森奶酪，面包糠。

shallot, parmesan cheese, breadcrumbs.

调料 Seasonings

芥花油，黄油，松露油，白糖，盐，黑胡椒，鸡蛋，蒜末，罗勒叶，柠檬汁，美乃滋。

canola oil, butter, truffle oil, white sugar, salt, black pepper, eggs, minced garlic, basil leaves, lemon juice, mayonnaise.

制作过程 Methods

1. 制作蘑菇丸子：将蘑菇切成粒状，圆葱切成末，净锅加入黄油、圆葱末、蒜末煸香，加入蘑菇继续煸炒，加入盐、黑胡椒、松露油调味。起锅凉透，打入鸡蛋、帕玛森奶酪、面包糠，做成小丸子。锅内倒入芥花油，炸至金黄色捞出。

   Make mushroom balls: Cut the mushrooms into grains, chop the shallot, add butter, minced shallot, minced garlic to stir-fry fragrant, add mushrooms continue to stir-fry, add salt, black pepper, truffle oil seasoning. Bring to a boil, whisk in eggs, parmesan cheese, breadcrumbs, and make small balls. Fry in canola oil until golden brown and remove.

2. 番茄洗净切片，拌入白糖、柠檬汁腌制2分钟。

   Rinse and slice tomatoes. Mix in sugar and lemon juice. Marinate for 2 minutes.

3. 将罗勒叶切成末状，和美乃滋搅拌均匀，作酱汁备用。

   Cut the basil leaves into final shapes. Stir well with mayonnaise, set aside for sauces.

4. 将蘑菇丸子、番茄片摆放盘内，适当浇淋美乃滋即可。

   Put the mushroom balls and tomato slices on the plate and pour over the mayonnaise.

技术解析 Technical resolution

制作丸子时，一定要等蘑菇凉透之后才可以放奶酪，否则奶酪容易融化流出，无法制作成品。

When making the mushroom balls, be sure to let the cooked mushrooms cool down before adding the cheese; otherwise, the cheese will melt out and it will not be possible to make the finished product.

十八、烘烤根茎菜和甜葱红椒配白色意大利香醋酱

No.18 Roasted Root Vegetables and Red Onions and Peppers with White Italian Balsamic Vinegar Sauce

小记 Notes

此菜是在中餐烹炒时蔬的基础上，转换成西餐烘烤的烹饪手法配以意大利香醋酱，使菜品保持原汁原味，更适合西方人的味蕾。

This dish is based on the concept of stir-fried vegetables in Chinese cuisine but is transformed using western-style baking techniques and Italian balsamic vinegar sauce, which makes the dish retain its original flavor and more suitable for western audiences.

主料 Main materials

红黄甜菜根，胡南瓜，小黄胡萝卜。

red and yellow beetroot, butternut squash, small yellow carrots.

配料 Ingredients

紫圆葱，红灯笼椒，小罗勒叶，绿芽。

purple shallot, red pepper, small basil leaves, green sprouts.

调料 Seasonings

橄榄油，黄油，意大利香醋，蜂蜜，盐，黑胡椒，帕玛森奶酪，芥末酱，干百里香，蒜末，蔬菜高汤。

olive oil, butter, Italian balsamic vinegar, honey, salt, black pepper, parmesan cheese, mustard sauce, dried thyme, minced garlic, vegetable stock.

制作过程 Methods

1. 将红黄甜菜根洗净，胡南瓜去皮切成方块，小黄胡萝卜一剖为二，紫圆葱和红灯笼椒切块。用锡纸包裹黄油、盐、红黄甜菜根，放入200℃的烤箱烤30分钟，取出红黄甜菜根，用干净的吸水纸擦去红黄甜菜根的表皮，然后切成块备用。

   Wash the red and yellow beetroot and cut the butternut squash into square pieces. Cut the small yellow carrots in half and cut the purple shallot and red peppers into pieces. Wrap butter, salt, red and yellow beetroot in foil and put them in a 200℃ oven for 30 minutes. Take out the red and yellow beetroot and wipe off the skin with clean absorbent paper towels, then cut them into pieces for later use.

2. 制作香醋酱汁：先将意大利香醋、蜂蜜、盐、黑胡椒、芥末酱、干百里香、蒜末混合搅拌均匀，然后再一边搅拌一边慢慢加入橄榄油，直到酱汁变黏稠时放入帕玛森奶酪，搅匀即可。

   Make balsamic vinegar sauce: First mix Italian balsamic vinegar, honey, salt, black pepper, mustard sauce, dried thyme and minced garlic evenly. Then slowly add olive oil while stirring until the sauce becomes thick. Add parmesan cheese while stirring.

3. 胡南瓜和小黄胡萝卜一起放入烤盘内加入黄油、盐、黑胡椒和少许蔬菜高汤放到200℃的烤箱烤15分钟。打开烤箱将甜菜根和紫圆葱块、红灯笼椒块放入盛有胡南瓜的烤盘中一起继续烤5分钟。从烤箱取出，放入适量的香醋酱汁搅拌均匀。

   Put butternut squash and small yellow carrots together in a baking pan with butter, salt, black pepper and a little vegetable stock and put them in a 200℃ oven for 15 minutes. Open the oven and put beetroot, purple shallot and red pepper into the butternut squash baking pan together for another 5 minutes of baking. Take it out of the oven and stir it with an appropriate amount of balsamic vinegar sauce.

4. 按照图片所示来装盘。用小罗勒叶和绿芽装饰。

   Follow the picture above to plate the dish. Garnish with small basil leaves and green sprouts.

技术解析 Technical resolution

1. 红黄甜菜根一定要带皮烤制，如果去皮或者切成块再烤，红黄甜菜根的颜色将改变，无法保持其本色。

   The red and yellow beetroot must be roasted with their skin on. If peeled or cut into pieces before roasting, the color of red and yellow beetroot will change permanently from their original color.

2. 酱汁中的橄榄油一定要慢慢地加入，太快加入无法达到黏稠的效果。

   Olive oil in the sauce must be added slowly. If added too quickly it will not achieve the desired thickness.

十九、焦香牛柳配香蕉酱、糖醋火腿、中式麦穗饺

No.19 Pan Seared Beef Tenderloin with a Buccaneer's Sauce, Sweet Sour Sausage and Chinese-style Dumpling

小记 Notes

此菜荣获2018年迪拜国际美食大赛金牌，是在单一的焦香牛柳的基础上，借鉴北美人喜爱的中餐糖醋味型和饺子理念，将焦香牛柳升级成为中西融合的一个代表菜。

This dish won the gold medal at the 2018 Dubai International Culinary Competition. It is a representative dish of Chinese-western fusion cuisine that upgrades the single grilled beef tenderloin by using the sweet and sour taste of Chinese food and the dumpling concept loved by North Americans.

主料 Main materials

牛柳，牛肉馅，牛培根。

beef tenderloin, beef stuffing, beef bacon.

配料 Ingredients

香蕉，胡萝卜，蟹味菇，西蓝花，彩虹甜菜根，紫甘蓝，欧防风根，肠衣，胡萝卜饺子皮，褐色牛汤。

banana, carrot, hypsizygus marmoreus, broccoli, rainbow beetroot, purple cabbage, parsnip, sausage casing, carrot dumpling wrapper, brown beef soup.

调料 Seasonings

芥花油，黄油，鸭油，白糖，草莓酱，盐，胡椒，奶油，葱末，姜末，蒜末，花雕酒，味极鲜酱油，鸡蛋，辣椒粉，圆葱末，番茄酱，醋，柠檬汁，面包糠，香芹末，白胡椒，百里香。

canola oil, butter, duck oil, white sugar, strawberry sauce, salt, pepper, cream, minced onion, minced ginger, minced garlic, Huadiao cooking wine, Weijixian soy sauce, eggs, chili powder, minced shallot, tomato sauce, vinegar, lemon juice, breadcrumbs, minced parsley, white pepper, thyme.

制作过程 Methods

❶ 将牛柳用牛培根卷起来放进真空袋里，然后放到63℃的恒温水槽中，煮35分钟，取出牛柳用黄油煎至牛柳表面呈金黄色。取一半裹匀面包糠和香芹末。

Roll the beef tenderloin with beef bacon and put it into a vacuum bag. Then put it into a constant water temperature of 63℃ and cook for 35 minutes. Take out the beef tenderloin and sear it with butter until the surface is golden brown. Coat half of it evenly with breadcrumbs and minced parsley.

❷ 将牛肉馅放入葱末、姜末、蒜末、花雕酒、味极鲜酱油、盐、胡椒、鸡蛋、辣椒粉搅拌均匀，取1/3的馅料做成麦穗饺子煮熟。将2/3的馅料灌入肠衣内，制作成小火腿肠，放入烤箱烤熟，锅内放入芥花油、蒜末炒香，加入番茄酱、白糖、醋调制成糖醋汁，将火腿肠裹匀糖醋汁即可。

Put minced onion, minced ginger and minced garlic into the beef stuffing. Add Huadiao cooking wine, Weijixian soy sauce, salt, pepper, eggs and chili powder to mix well. Make 1/3 of the stuffing into wheat ear dumplings and cook them. Inject 2/3 of the stuffing into sausage casing to make small sausages. Roast in the oven until cooked. Put canola oil in the pan and stir-fry minced garlic until fragrant. Add tomato sauce, white sugar and vinegar to make sweet and sour sauce. Coat the sausage with sweet and sour sauce.

❸ 将胡萝卜、西蓝花、蟹味菇洗净焯水，锅内放入黄油、圆葱末炒香，放入蔬菜、盐、胡椒调味。

Carrots, broccoli and mushrooms are washed and blanched in boiling water. Put butter and minced shallot in the pan and stir-fry until fragrant. Add vegetables, as well as salt and pepper to season.

❹ 将紫甘蓝切丝，用鸭油煸炒，加入草莓酱、柠檬汁煨至紫甘蓝软糯。

Cut purple cabbage into shreds and stir-fry with duck oil. Add strawberry sauce and lemon juice and simmer until purple cabbage is soft.

❺ 制作欧防风根泥：将欧防风根去皮切成小块，放入锅内加入黄油、奶油、盐、白胡椒、百里香用小火加热至锅内奶油剩下1/3，离火倒入搅拌机内，搅拌至细腻没有颗粒。

Make parsnip puree: Peel and wash the parsnip, cut into small pieces and put it into a pot with butter, cream, salt, white pepper and thyme. Heat over low heat until there is only 1/3 of cream left in the pot. Pour it into a blender and blend until pureed.

6. 制作酱汁：净锅放入褐色牛汤、香蕉块，小火煨至汤略微黏稠，过滤后加入冷黄油搅拌均匀。

   Make sauce: Put brown beef soup and banana pieces in a saucepan. Simmer over low heat until the sauce is slightly thickened. Strain and add cold butter to stir well.

7. 按图所示装盘，用彩虹甜菜根装饰即可。

   Fill the plate according to the drawing, decorate with rainbow beetroot.

技术解析 Technical resolution

欧防风根泥不宜长时间搅拌。酱汁要放入冷黄油来增加酱汁的黏稠度。

Parsnip puree should not be blended for a long time. Cold butter should be added to increase the thickness of the sauce.

二十、低温慢煮三文鱼和比目鱼

No.20 Sous Vide Salmon and Halibut

小记 Notes

加拿大卑诗省盛产三文鱼和比目鱼，因此三文鱼和比目鱼在加拿大人的餐桌上是常见的，采用低温慢煮的烹调方法保持了三文鱼本身的鲜味与嫩度。酱汁的搭配采用中式酱香味型，用中式的蜂巢包裹着西式鱼肉浓汤。

British Columbia is famous for its salmon and halibut, so these fishes are very common on Canadian tables. The sous vide method preserves the freshness and tenderness of the fish itself. The sauce is paired with Chinese-style soybean paste sauce, while the western-style fish soup is wrapped in a Chinese-style honeycomb.

主料 Main materials

三文鱼，比目鱼。

salmon, halibut.

配料 Ingredients

虾仁，水萝卜，芦笋，西芹粒，胡萝卜粒，红椒粒。

shrimp, red radish, asparagus, diced celery, diced carrot, diced red pepper.

调料 Seasonings

芥花油，圆葱，姜，大料，香叶，百里香，鲜奶油，盐，白糖，酱油，甜面酱，料酒，胡椒粉，白葡萄酒，鸡蛋清，黄油，面包糠，香芹，松仁，澄粉，凝胶片，鱼汤。

canola oil, shallot, ginger, star anise, bay leaves, thyme, fresh cream, salt, white sugar, soy sauce, sweet soybean paste, cooking wine, pepper powder, white wine, egg white, butter, breadcrumbs, parsley, pine nuts, wheat starch, soaked gelatin sheet, fish soup.

制作过程 Methods

❶ 将三文鱼去骨去皮片成1厘米厚、10厘米宽的长方形片，比目鱼去骨去皮切成长条，保留边角料备用。

Remove the bones and skin of the salmon and cut it into rectangular slices 1 cm thick and 10 cm wide. Remove the bones and skin of the halibut and cut it into long strips. Reserve the trimmings.

❷ 将2/3的虾仁剁碎成泥，加入剩下1/3的虾仁、西芹粒、红椒粒、盐、鸡蛋清、胡椒粉搅拌均匀成虾胶备用。

Chop 2/3 of the shrimps into a paste, add the remaining 1/3 of the shrimps, diced celery, diced red pepper, salt, egg white, pepper powder and stir evenly into shrimp paste.

❸ 平铺保鲜膜，把三文鱼放上去，然后把虾胶放在三文鱼上面，再将比目鱼放在虾胶上面，卷成半圆形，装入密封袋子里放入50℃的恒温水槽中煮40分钟，捞出放入冰水中过凉，捞出切成需要的大小，放入冷藏室备用。

Flat the saran wraps and place the salmon on it. Then put the shrimp paste on top of the salmon then put the halibut on top of the shrimp paste. Roll into a semicircle. Put it in a sealed bag and put it in a 50°C sous vide for 40 minutes. Take it out and put it in ice water. Cut it into the required size as needed and store it in the fridge.

❹ 将黄油、面包糠、香芹、松仁放入搅拌机内搅碎成细沙状，倒在吸油纸上，上面再铺一层吸油纸，用擀面杖擀成薄薄的一层，放入冷冻室20分钟后取出切成长方形状，铺在之前切好的鱼上。将其放入120℃的烤箱内，加热至内部温度达到120℃。

Put butter, breadcrumbs, parsley and pine nuts into a blender and blend until fine sand-like. Pour it on parchment paper and then cover it with another parchment paper on top. Roll it into a thin layer with a roller. After 20 minutes in the freezer, cut it into rectangular shapes and place them on the previously cut fish. Put them in a 120 °C oven and heat them until the internal temperature reaches 120 °C.

❺ 制作鱼肉浓汤球：锅内放入黄油、圆葱粒、西芹粒、胡萝卜粒煸香之后，加入鱼汤、白葡萄酒、香叶、百里香，大火烧开，放入鱼肉的边角料，转小火煨制，将香叶、百里香取出，加入鲜奶油、盐、胡椒粉调味，放入化好的凝胶片搅拌均匀，倒入5克小圆形硅胶模具，放入冷藏室备用。

Make fish soup ball: Put butter, diced shallot, diced celery and diced carrot in a pot and sauté until soft. Add fish soup, white wine, bay leaves and thyme on high heat and boil. Then add the fish trimmings and simmer on low heat. Remove bay leaves and thyme, then add fresh cream, salt and pepper powder to taste and add soaked gelatin sheet. Stir well. Pour into 5g small round silicone molds and refrigerate for later use.

❻ 制作蜂巢：和面机内放入澄粉、盐，加入开水搅拌均匀，再加入黄油，将油和面团混合均匀，取出保鲜膜包住放入冷藏室冷却30分钟，取出分成每个15克的小剂子，每个剂子用拇指按出一个坑将鱼肉浓汤球包进去，放入180℃的油锅中炸制2分钟，捞出控油即可。

Make honeycomb: Put wheat starch and salt in a bowl, add boiling water to stir evenly then add butter to knead the dough until the butter and dough mix well. Wrap with saran wrap and refrigerate for 30 minutes. Take it out and divide into small doses of 15g each. Use your thumb to press out a hole for each dose to wrap the fish soup ball, then slowly put it in a 180℃ deep fryer for 2 minutes. Remove and control oil.

❼ 水萝卜和芦笋分别放入装有融化的黄油袋子里，撒入盐和胡椒粉密封，放入蒸箱蒸熟即可。

Red radish and asparagus are put into bags with melted butter, sprinkled salt and pepper powder on top, sealed and steamed in a steamer.

❽ 制作酱汁：锅内放入芥花油，加入圆葱、姜、大料、甜面酱煸香之后，加入酱油、水、料酒、白糖、鱼骨大火烧开，撇去浮沫，小火熬制直到汤汁黏稠，过滤取汁即可。

Make sauce: Add canola oil to the pot, then add shallot, ginger, star anise, sweet soybean paste and stir-fry until fragrant. Add soy sauce, water, cooking wine, white sugar and fish bones. Boil over high heat. Skim off the foam and simmer over low heat until the soup is thickened. Strain out the sauce.

❾ 装盘：将鱼放在盘子的左侧，蜂巢摆在盘子的右上角，水萝卜和芦笋放在鱼和蜂巢中前方，靠近鱼的内侧淋入酱汁。

Plate：Place the fish on the left side of the plate, followed by the honeycomb on the upper right corner of the plate, then the red radish and asparagus in front of the fish and honeycomb, respectively. Lastly, pour sauce on top of the fish.

技术解析 Technical resolution

❶ 三文鱼和比目鱼一定要剔净骨刺和外皮。

Salmon and halibut must be removed bone and skin.

❷ 取出低温慢煮的三文鱼、虾胶、比目鱼卷后先用冰水浸泡，然后切成需要的大小后再放入冰箱中冷藏。

Take out sous vide cooked salmon, shrimp, halibut roll and put it in ice water first, then cut it into the required size as needed and store it in the fridge.

二十一、冰皮酸辣黑豆豉鮟鱇鱼
No.21 Snow Skin Hot and Sour Black Bean Monkfish

小记 Notes

此菜是中西式结合的一个冷前菜，是通过加入中式点心中的冰皮和加拿大人喜欢的中式黑豆豉鮟鱇鱼的做法改良而来的。

This dish is a cold appetizer that combines Chinese and western cuisine. It is created by using the idea of the Chinese snow skin mooncake, with the addition of fried monkfish with salted black beans that Canadians love.

主料 Main materials

鮟鱇鱼。

monkfish.

配料 Ingredients

酸黄瓜，番茄嫩叶。

pickled cucumber, tender leaves of tomatoes.

调料 Seasonings

芥花油，葱，姜，蒜，大料，酱油，花雕酒，黑豆豉，番茄酱，鸡蛋清，辣椒粉，香芹末，琼胶，牛奶，白糖，盐，糯米粉，粘米粉，澄粉，泰式酸甜辣汁。

canola oil, shallot, ginger, garlic, star anise, soy sauce, Huadiao cooking wine, black bean, tomato sauce, egg white, chili powder, minced parsley, agar, milk, white sugar, salt, glutinous rice flour, sticky rice flour, wheat starch, Thai sweet and spicy sauce.

制作过程 Methods

❶ 将鮟鱇鱼洗净去皮去骨留肉。将鮟鱇鱼尾部的肉用保鲜膜卷紧成圆形，真空包装，放入冷冻室冷冻30分钟。

Wash the monkfish and remove the skin and bones. Roll the meat into a round shape with saran wrap and vacuum pack it for 30 minutes in the freezer.

❷ 将剩余的鱼肉分成两份，一份加工成鱼胶，另一份做成黑豆豉鱼肉。

Divide the remaining fish meat into two parts. One part is processed into fish paste and the other part is made into black bean sauce fish meat.

❸ 将鱼肉剁碎成泥，加入鸡蛋清、盐、番茄酱、辣椒粉、香芹末搅拌均匀成鱼胶备用。

Chop the fish meat into ground fish and mix it with egg white, salt, tomato sauce, chili powder and minced parsley to make fish paste for later use.

❹ 锅内放入芥花油加热至八成热，放入鱼肉炸干，捞出控油。锅内留少许油爆香葱、姜、蒜、大料、黑豆豉，加入水、酱油和花雕酒烧开，加盐调味，放入炸好的鱼肉，小火煨至软嫩。捞出平铺在不锈钢平盘内，上面放重物将鱼肉压平，切成需要的长方形。

Put canola oil in a pan and heat it up until it reaches 180°C. Deep fry it until dry. Leave a little oil in the pan to sauté the shallot, ginger, garlic, star anise and black bean. Add water, soy sauce and Huadiao cooking wine to boil. Season with salt and add fried fish meat to simmer over low heat until it becomes soft and tender. Take out the fish meat and spread it flat on a hotel pan. Put heavy weight on top of it to press it flat and cut it into rectangular pieces as needed.

❺ 将酸黄瓜切成细小颗粒备用。将煨制鱼肉的黑豆豉汁过滤无渣后倒入锅中小火加热，加入酸黄瓜粒、辣椒粉调成酸辣味型，放入琼胶搅拌均匀，倒入平盘内放入冷藏室冷却，切成需要的长方形。

Cut pickled cucumber into small pieces for later use. Strain from black bean soup which is used for simmering fish meat and pour it into a pot to heat up over low heat. Add pickled cucumber pieces and chili powder. Stir in agar until completely dissolved. Pour it into a flat plate for cooling in fridge until firm, then cut into rectangular pieces as needed.

❻ 制作冰皮：将芥花油、牛奶、盐、糯米粉、粘米粉、澄粉、白糖搅拌均匀至无颗粒，放入蒸锅蒸30分钟，取出凉凉，用手揉匀，擀成薄饼状，放入热锅煎至两面略黄即可。

Make snow skin: Mix canola oil, milk, salt, glutinous rice flour, sticky rice flour, wheat starch and white sugar together without lumps. Steam for 30 minutes in a steamer then take it out to cool down before kneading by hand into dough for later use. Roll out thin-like pastry dough then sear in a hot pan until slightly browned on both sides.

❼ 将保鲜膜平铺在方形的模子内，将冰皮铺在模子的底部和两侧，挤入鱼胶，再将卷好的鱼卷放入鱼胶上，用鱼胶填补空隙抹平放上压平的鱼肉，然后用保鲜膜裹紧，装入密封袋子里放入50℃的恒温水槽中30分钟。取出打开模子在上方刷一层黑豆豉汁然后将酸黄瓜啫喱铺在上面，保鲜膜裹紧后放入冰箱冷藏至内部温度到0℃。取出切成1.5厘米厚的片。

Lay saran wraps flat inside a square mold, then lay snow skin on the bottom and sides of the mold before squeezing in the fish paste. Then place rolled-up fish on top of the fish paste. Fill gaps with fish paste then make the paste flat. Wrap tightly with saran wrap and seal it for cooking in 50℃ sous vide for 30 minutes. Take out the mold then brush black bean sauce on top before laying pickled cucumber jelly on top of that. Then wrap tightly with plastic wrap again before refrigerating until internal temperature reaches 0℃. Cut into slices that are 1.5 cm thick.

❽ 装盘：将鱼片平铺盘子中间，插入番茄嫩叶，淋入泰式酸甜辣汁。

Plate: Lay fish slices flat in the middle of the plate, then insert tender leaves of tomatoes, and pour Thai sweet and spicy sauce.

技术解析 Technical resolution

❶ 鮟鱇鱼必须去除骨刺和皮。

Monkfish must have bone spurs and skin removed.

❷ 鱼胶卷要在规定的温度内进行冷藏处理。

The fish film should be refrigerated at the specified temperature.

二十二、生腌三文鱼佐食青萝卜花生酱
No.22 Cured Salmon with Peanut Sauce Green Radish

小记 Notes

此菜是一道比较清淡的前菜。运用最简单的盐糖腌制方法腌制新鲜的三文鱼，青萝卜蘸花生酱是山东胶东半岛当地人比较喜爱的一个蘸酱菜，将两者搭配起来作为一道西式的前菜，口感相互协调。

This dish is a light appetizer. The freshest salmon is pickled using the simplest salt and sugar-cured method. Green radish dipped in peanut sauce is a popular local food in Jiaodong Peninsula, Shandong Province. The two are paired together as a western-style appetizer with complementary flavors.

主料 Main materials

大西洋三文鱼。

Atlantic salmon.

配料 Ingredients

青萝卜，水萝卜皮，樱桃花。

green radish, red radish skin, cherry blossom.

调料 Seasonings

盐，白糖，樱桃，柠檬汁，去皮花生，细辣椒粉，醋。

salt, white sugar, cherries, lemon juice, peeled peanuts, fine chili powder, vinegar.

制作过程 Methods

❶ 将刺身级别的三文鱼宰杀去皮洗净，切成长条，用1：1的盐和糖混合物腌制60分钟，冷水冲洗干净，放入1：1的水和醋混合液中浸泡1分钟，捞出吸干水分，用保鲜膜卷紧成圆形，放入冷藏室内备用。

Fillet the sashimi-grade salmon, remove the skin, and wash it clean. Cut it into long strips, cure it with a mixture of salt and sugar in a ratio of 1：1 for 60 minutes, then rinse it clean with cold water. Put it into a mixture of water and vinegar in a ratio of 1：1 for 1 minute, remove it and pat dry. Wrap tightly with saran wrap into a round shape and put it in the fridge for later use.

❷ 将青萝卜洗净，用刨片机刨成长条薄片，放入开水锅中焯水至熟，捞出过凉控水卷成圆筒状。将水萝卜皮切成丝竖着插入萝卜卷中心。

Wash the green radish clean, slice it into long thin slices with a slicer, and blanch it in boiling water until cooked. Remove it and cool it down with cold water, then roll it into a cylinder shape. Cut the red radish skin into julienne and insert them vertically into the center of the radish roll.

❸ 制作樱桃酱：锅内放入去核的樱桃、白糖、水、柠檬汁，大火烧开后转小火熬至黏稠，然后用搅拌机搅拌均匀，装入盛器凉凉。

Make cherry sauce: Put pitted cherries, white sugar, water and lemon juice in a pot. Boil over high heat and simmer until thickened, then blend evenly with a blender and put it into a container to cool.

❹ 自制花生酱：将烘烤的去皮花生放入搅拌机内，倒入适量的纯净水、盐，搅拌至均匀细腻无颗粒状，加入细辣椒粉搅拌均匀即可。

Homemade peanut sauce: Put roast peeled peanuts into a blender, then add an appropriate amount of pure water and salt. Blend them evenly without lumps, add fine chili powder and mix well.

❺ 装盘：腌制好的三文鱼卷切成0.5厘米厚的圆片，放入盘中，三文鱼上面再放上少许樱桃酱，用樱桃花点缀。摆上青萝卜卷，然后在中间放入自制花生酱。

Plate: Cut the pickled salmon roll into 0.5 cm round slices and place them on a plate. Put some cherry sauce on top of the salmon and garnish with cherry blossom. Place the green radish roll on top of it and then put home made peanut sauce in the middle.

技术解析 Technical resolution

三文鱼必须用盐、糖混合物腌制后再用水和醋混合液浸泡，以杀死生鱼表面细菌。

The salmon must be cured with a mixture of salt and sugar and then soaked in a mixture of water and vinegar to kill bacteria on the surface of the raw fish. .

二十三、茶香三文鱼鳕鱼配无花果沙拉

No.23 Green Tea Salmon and Cod with Fig Salad

小记 Notes

此菜运用中国的绿茶，将三文鱼和鳕鱼用绿茶水加上辅助材料腌制入味，再进行低温蒸制，突出淡淡的绿茶香味。

This dish uses Chinese green tea and auxiliary ingredients to marinate the salmon and cod. Steaming this dish at a low temperature will subsequently highlight the light tea flavor.

主料 Main materials

大西洋三文鱼，鳕鱼。

Atlantic salmon, cod.

配料 Ingredients

无花果，时蔬，三文鱼子。

fig, seasonal vegetable, salmon roe.

调料 Seasonings

绿茶水，大料，茴香，盐，白糖，绿茶油，白芝麻，酱油，圆葱碎，日本米醋，黑胡椒粉。

green tea water, star anise, fennel, salt, white sugar, green tea oil, white sesame, soy sauce, chopped shallot, Japanese rice vinegar, black pepper powder.

制作过程 Methods

❶ 将绿茶水、茴香、大料、盐、白糖放入一个大的不锈钢锅中，小火加热混合搅拌均匀，直到白糖和盐溶化。

Put green tea water, fennel, star anise, salt and white sugar into a pot and heat to a boil. Stir until the white sugar and salt are melted.

❷ 将三文鱼和鳕鱼宰杀去皮洗净，切成长条，放入绿茶水溶液中，液体需要漫过三文鱼，浸制90分钟。捞出，吸干水分。

Fillet the salmon and cod, remove their skin, and wash them. Cut them into long strips and put them in the green tea brine. Marinate the salmon for 90 minutes. Take them out and pat dry.

❸ 将三文鱼片成厚度均匀的长方形片，将鳕鱼条包在里面，用保鲜膜卷紧成圆柱形，真空包装，放入50℃的蒸箱内蒸1小时，捞出连真空袋一起放入冰水中过凉，直至内部温度降至0℃，再横切成5厘米长的片，然后纵向切成两个相等的半圆形。

Butterfly the salmon into evenly thick rectangular pieces and wrap the cod strips inside. Roll it tightly into a circle with saran wrap. Vacuum and seal. Steam it at a 50°C steamer for one hour. Take it out and put it in ice water until the internal temperature is 0°C. Crosscut it into 5 cm lengths then cut lengthwise into two equal halves with a sharp knife.

4. 将无花果切成橘子瓣形备用。时蔬菜洗净。

   Cut the fig into petals for later use. Wash the seasonal vegetable.

5. 制作茶香芝麻沙拉酱：将绿茶油、白芝麻、酱油、圆葱碎、日本米醋放入搅拌机内搅碎均匀至没有颗粒状，放入黑胡椒粉拌匀即可。

   Make tea-flavored sesame salad dressing: Put green tea oil, white sesame, soy sauce, chopped shallot and Japanese rice vinegar into a blender and blend until smooth without particles. Add black pepper powder and mix well.

6. 装盘：将蒸好过凉的三文鱼鳕鱼卷顺刀切成均匀的两半，取一半装盘，上面放上三文鱼子。再放入时蔬和无花果，淋上酱汁即可。

   Plate: Cut the steamed salmon cod roll into two halves. Plate half of the roll and top with the salmon roe. Add seasonal vegetable and fig and drizzle with the sauce.

技术解析 Technical resolution

三文鱼和鳕鱼宰杀去皮洗净后，要放入绿茶水中漫泡浸制90分钟。

Salmon and cod are butchered, peeled and washed, then soaked in green tea water for 90 minutes.

二十四、北极三文鱼和野生三文鱼配卡夫奇妙酱
No.24 Arctic Salmon and Wild Salmon with Kraft Miracle Whip

小记 Notes

此菜使用加拿大盛产的北极三文鱼和野生三文鱼搭配卡夫奇妙酱。

This dish is made with Canadian Arctic salmon and wild salmon served with kraft miracle whip.

主料 Main materials

北极三文鱼，野生三文鱼。

Arctic salmon, wild salmon.

配料 Ingredients

新鲜的腐竹皮。

fresh bean curd skin.

调料 Seasonings

盐，白糖，漆树粉，香芹末，卡夫奇妙酱，芥末油。

salt, white sugar, sumac powder, minced parsley, kraft miracle whip, mustard oil.

制作过程 Methods

❶ 将北极三文鱼宰杀去皮洗净取肉，片成厚度均匀的大片，用白糖、盐腌制1小时，洗净吸干水分备用。

Fillet the Arctic salmon and remove the skin and wash. Cut it into evenly thick pieces and cure it with white sugar and salt for 1 hour. Rinse it with cold water and pat dry.

❷ 将野生三文鱼宰杀去皮洗净，切成长条，用白糖、盐、芥末油腌制1小时，洗净吸干水分备用。

Fillet the wild salmon and remove the skin and wash. Cut it into long strips and cure it with white sugar, salt and mustard oil for 1 hour. Rinse it with cold water and pat dry.

❸ 用北极三文鱼将野生三文鱼条包在里面，用保鲜膜卷紧成自然形状，真空包装，放入40℃的恒温水槽中40分钟，捞出连真空袋一起放入冰水中过凉，直至内部温度降至0℃。捞出切成1厘米的厚片备用。

Place the wild salmon strips in Arctic salmon, roll them tightly into a natural shape with saran wrap, vacuum pack them and seal. Put them in a sous vide at 40°C for 40 minutes. Take it out and put in the ice water until the internal temperature of fish is 0°C. Take out and cut them into 1 cm thick slices for later use.

❹ 将新鲜的腐竹皮切成大小符合要求的长方形，放在两个半圆形的模子中夹紧，放入八成热的油温中炸至酥脆。

Cut fresh bean curd skin into rectangular shapes of required size, put them between two semi round molds, fry them at 180°C oil until crispy.

❺ 装盘：将切好的鱼肉平放到盘中，淋上几滴芥末油，将炸好的腐竹皮放在鱼肉上，挤入卡夫奇妙酱，撒上漆树粉和香芹末。

Plate: Place the fish flat on a plate, drizzle a few drops of mustard oil on top of it, then place the fried bean curd skin on top of the fish and squeeze in kraft miracle whip, sprinkle with sumac powder and minced parsley.

技术解析 Technical resolution

❶ 北极三文鱼宰杀取肉，要用白糖、盐腌制1小时备用。

Cured the Arctic salmon with white sugar and salt for 1 hour.

❷ 野生三文鱼宰杀取肉后要用白糖、盐、芥末油腌制1小时，洗净吸干水分备用。

Cured the wild salmon with white sugar, salt, mustard oil for 1 hour, rinse and drain the water reserve.

二十五、芝麻辣酱青萝卜海鲜

No.25 Green Radish Seafood Press with Spicy Sesame Sauce

小记 Notes

此菜是在罗汉肚的基础上改良而来的，用青萝卜皮代替猪肚包裹各种海鲜再配以芝麻辣酱而成。

This dish is created based on the concept of Chinese Luo Han Du. Using green radish skin to wrap all seafoods and press, then served with spicy sesame sauce.

主料 Main materials

老板鱼，鱿鱼，三文鱼，鲈鱼，虾仁。

skate, squid, salmon, perch, shrimp.

配料 Ingredients

黄瓜，青萝卜，腌制的橄榄果。

cucumber, green radish, pickled olives.

调料 Seasonings

橄榄油，酱油，芥末酱，盐，柠檬汁，糖，黑胡椒，胡椒粉，老干妈香辣酱，熟白芝麻，鸡蛋清。

olive oil, soy sauce, mustard sauce, salt, lemon juice, sugar, black pepper, pepper powder, spicy pepper paste, roasted white sesame, egg white.

制作过程 Methods

❶ 将老板鱼去骨去皮，鱿鱼洗净切成条状，鲈鱼和三文鱼宰杀去骨去皮切条，用糖、盐和芥末酱腌制1小时。冲洗干净，吸干水分备用。

Remove the bones and skin from the skate and wash. Wash the squid and cut it into strips. Remove the bones and skin from the perch and salmon and cut them into strips. Cure them with sugar, salt and mustard sauce for 1 hour. Rinse them clean and pat dry.

❷ 取青萝卜皮备用，锅内烧开水放入萝卜皮略煮捞出过凉，放入真空袋中，再放入盐和糖，真空放入冷藏室备用。

Place the green radish skin in the boiling water. Take it out and cool it down. Put it in a vacuum bag with salt and sugar. Vacuum seal it and place in the fridge for later use.

❸ 将虾仁剁碎成泥，放入芥末酱、盐、鸡蛋清、胡椒粉、柠檬汁搅拌均匀成虾胶备用。将处理好的老板鱼、鱿鱼、鲈鱼和三文鱼和虾胶搅拌均匀。保鲜膜上铺上青萝卜皮然后抹一层薄薄的虾胶再

平铺老板鱼肉，然后在老板鱼肉上依次摆上鱿鱼、鲈鱼、三文鱼和黄瓜条。然后用保鲜膜卷紧成圆筒形。在两侧各放一块铁板，放入真空袋中，使两块铁板之间压紧实，放入65℃的蒸箱中蒸20分钟取出。放入冰水中使内部温度降到0℃。取出切成3厘米的块备用。

Chop the shrimp into paste and mix it with mustard sauce, salt, egg white, pepper powder and lemon juice to make shrimp paste for later use. Mix the processed skate, squid, perch and salmon with shrimp paste. Spread green radish skin on saran wrap and spread a thin layer of shrimp paste on top of it. Then lay a flat skate on top of it. Place squid, perch, salmon and cucumber on top of skate. Then roll it tightly into a cylinder shape with saran wrap. Put an iron plate on each side and vacuum pack it between two iron plates to compress it tightly. Put it in a 65℃ steamer for 20 minutes and take it out. Put it in ice water until the internal temperature drops to 0℃. Remove and cut into 3cm pieces. Set aside.

4. 制作芝麻辣酱：破壁机内放入橄榄油、酱油、熟白芝麻、老干妈香辣酱、柠檬汁、糖、黑胡椒搅碎均匀至无颗粒状即可。

   Make spicy sesame sauce：Put olive oil, soy sauce, roasted white sesame, spicy pepper paste, lemon juice, sugar and black pepper in a blender until smooth.

5. 装盘：盘中间放入酱汁，用瓶底压制酱汁，然后放入鱼块，在鱼块上放上腌制好的橄榄果和少许的老干妈香辣酱即可。

   Plate: Place the sauce in the middle of the plate and press down on the bottom of the bottle to release the sauce, then place fish pieces on top of the sauce and put pickled olives, a little spicy pepper paste on top of fish.

技术解析 Technical resolution

老板鱼肉、鱿鱼肉、鲈鱼肉和三文鱼肉，要用糖、盐、芥末酱腌制1小时。

Skate, squid, perch and salmon, marinated in sugar, salt and mustard sauce for 1 hour.

二十六、海苔扇贝柱佐食橄榄菜
No.26 Nori Scallop with Olive Vegetable

小记 Notes

此菜是将潮州菜的小吃橄榄菜与扇贝柱结合起来，运用中式橄榄菜搭配西式扇贝柱的烹调方法制成。

This dish combines the Chaozhou snack olive vegetable with scallop slices, using the cooking method of mixing Chinese olive vegetable with western-style scallop slices.

主料 Main materials

扇贝柱。

scallop.

配料 Ingredients

芦笋，水萝卜，柠檬，玫瑰花瓣。

asparagus, red radish, lemon, rose petals.

调料 Seasonings

黄油，美乃滋，炼乳，豆酱，海苔，橄榄菜，盐，粘肉粉，白糖，黑胡椒粉。

butter, mayonnaise, condensed milk, bean paste, nori, olive vegetable, salt, meat glue powder, white sugar, black pepper powder.

制作过程 Methods

❶ 将扇贝柱洗净控干水分，撒入盐、黑胡椒粉、粘肉粉拌匀，然后将扇贝柱依次横放摆在海苔上，用海苔包住扇贝柱，用保鲜膜卷成圆筒形，放入50℃的恒温水槽中30分钟。取出，然后拿掉扇贝柱表面的海苔，切成2.5厘米长的段备用。

Wash the scallop and dry them. Sprinkle salt, black pepper powder and meat glue powder evenly on the scallop. Place the scallop horizontally on the nori one by one. Wrap the scallop with nori and roll them into a cylinder shape with saran wrap. Put them in a 50°C water tank for 30 minutes. Take it out and remove the nori from the surface of the scallop. Cut them into 2.5 cm long sections for later use.

❷ 将水萝卜切成片，用白糖、柠檬汁、盐拌匀备用，芦笋切片焯盐水备用。

Cut red radish into slices and marinate with white sugar, lemon juice and salt for later use. Cut asparagus into slices and blanch in salt water for later use.

❸ 将美乃滋、炼乳、豆酱倒入一个大碗中搅拌均匀备用。将橄榄菜、黄油放入搅拌机内，搅至均匀至黏稠状即可。

Pour mayonnaise, condensed milk and bean paste into a large bowl and mix well for later use. Put olive vegetable and butter in a blender and blend until evenly thickened.

❹ 装盘：将水萝卜铺在盘子上面，依次摆上扇贝柱、芦笋片、橄榄菜，再将混合的美乃滋交叉挤在扇贝柱的两侧，上面用玫瑰花瓣点缀。

plate: Place red radish on the plate and then place scallop, asparagus and olive vegetable one by one. Then squeeze the mixed mayonnaise on both sides of the scallop. Garnish with rose petals on top.

技术解析 Technical resolution

❶ 扇贝柱要用盐、黑胡椒粉腌制入味。

The scallop should be marinated with salt and black pepper powder.

❷ 扇贝柱用海苔包住，恒温30分钟后取出，将扇贝柱表面的海苔去掉。

Scallop with nori wrap, constant temperature after 30 minutes remove without nori.

二十七、鸡翅炖鲍鱼
No.27 Braised Abalone with Chicken Wings

小记 Notes

鲍鱼在加拿大人的餐桌上是很少见的，但是鸡翅则是常见的。酱油是加拿大人比较喜欢的调味品之一，因此这道菜用中式酱油煮的烹调方法将鲍鱼和鸡翅放在一起烹制，让彼此香味互补，突出酱香味，让西方人更容易接受鲍鱼。成品口味咸鲜微甜，酱香浓郁。

Abalone is rare on Canadian tables, but chicken wings are common. Soy sauce is one of the seasonings that Canadians like, so this dish is created by using Chinese soy sauce to cook abalone along with chicken wings. Both of ingredients complement each other's flavor and highlight the soy sauce flavor, making it easier for westerners to accept abalone. The dish tastes salty, fresh, slightly sweet, with a strong soy sauce flavor.

主料 Main materials

鲍鱼，鸡翅。

abalone, chicken wings.

配料 Ingredients

奶油白菜。

bok choy.

调料 Seasonings

芥花油，葱，姜，蒜，盐，白糖，酱油，香卤汁，料酒，干辣椒，鸡汤。

canola oil, green onion, ginger, garlic, salt, white sugar, soy sauce, Chinese marinade, cooking wine, dried chili pepper, chicken broth.

制作过程 Methods

1. 将鸡翅焯水洗净备用。

   Blanch the chicken wings and wash them for later use.

2. 将鲍鱼洗净去壳，去内脏，剞上十字花刀，焯水备用。

   Wash the abalone and remove the shell and internal organs. Crosscut and blanch it for later use.

3. 锅内放入芥花油，爆香葱、姜、蒜、干辣椒，加入鸡汤、酱油、香卤汁、料酒，放入鸡翅和鲍鱼，加入盐、白糖调味，大火烧开，转小火炖到鸡翅和鲍鱼煮熟，取出。小火煨至汤汁黏稠成酱汁。

   Put canola oil in the pot and sauté green onion, ginger, garlic and dried chili pepper until fragrant. Add chicken broth, soy sauce, Chinese marinade, cooking wine, chicken wings and abalone into the pot. Add salt and white sugar to taste and bring to a boil over high heat and simmer over low heat until the meat is cooked. Pull out the meat and continue to reduce the soup until slightly thickened. Strain to become a sauce.

4. 将奶油白菜洗净，一切为二，放入盐水中煮至断生捞出备用。

   Wash the bok choy and cut it in half. Put it in salt water and cook until it is just cooked. Take it out and set it aside.

5. 装盘：将鲍鱼和鸡翅各取一只装到盘子中间，外侧放上奶油白菜，淋入煨制鲍鱼鸡翅时的酱汁即可。

   Plate: Take one abalone and one chicken wing and put them in the middle of the plate. Put bok choy on the outside and pour in the sauce.

技术解析 Technical resolution

鲍鱼焯水以去除腥味为主，时间要短。

Abalone boiled water to remove the main fishy smell, the time to be short.

二十八、龙虾扇贝球佐食车达奶酪西蓝花汤

No.28 Lobster Scallop ball with Cheddar Cheese Broccoli Soup

小记 Notes

加拿大的龙虾和扇贝是北美人比较喜欢吃的海鲜。此菜运用中餐制作虾胶的技法将龙虾制作成龙虾胶卷起扇贝柱低温慢煮，再配以西餐中的车达奶酪西蓝花汤而成。

Canadian lobster and scallop are seafood that North Americans like to eat. This dish is prepared by using Chinese cooking techniques to make lobster paste, adding more flavor into the dish. Rolled scallop with lobster paste are slowly cooked at a low temperature and served with the western food of cheddar cheese broccoli soup.

主料 Main materials

龙虾尾，扇贝柱。

lobster tail, scallop.

配料 Ingredients

西蓝花，白蘑菇，青豆，圆葱粒，西芹粒。

broccoli, white mushrooms, green peas, diced shallot, diced celery.

调料 Seasonings

黄油，鸡蛋清，花雕酒，葱姜汁，盐，白胡椒粉，面粉，车达奶酪，奶油，牛奶，鸡汤。

butter, egg white, Huadiao cooking wine, onion and ginger juice, salt, white pepper powder, flour, cheddar cheese, cream, milk, chicken broth.

制作过程 Methods

❶ 将龙虾尾去壳，取龙虾肉，留龙虾皮备用。龙虾肉加入葱姜汁、花雕酒、鸡蛋清、白胡椒粉、盐调均匀，用搅拌机打成虾胶取出。将龙虾皮平铺在保鲜膜上，再将龙虾胶均匀铺在龙虾皮上，然后放上扇贝柱，用保鲜膜卷起成球状，放入50℃的水槽中慢煮25分钟。

Remove the shell from the lobster tail, take off the lobster meat and save the lobster skin for later use. Add lobster meat, onion and ginger juice, Huadiao cooking wine, egg white, white pepper powder and salt to the container of the food processor. Process until pureed. Spread the lobster skin evenly on the saran wrap, spread the lobster paste evenly on the lobster skin, then put the scallop in the middle, roll it into a ball with the saran wrap, put it in a water tank at 50℃ and cook for 25 minutes.

❷ 将西蓝花、白蘑菇和青豆焯水，留出一部分装饰用，其余的粗略切碎，锅中加入黄油用中火加热，加入圆葱粒、西芹粒、白蘑菇、西蓝花和青豆煸炒，加入面粉继续搅拌至黏稠，慢慢加入鸡汤，加热烧开，不停地搅拌，直到黏稠，倒入搅拌机内搅拌成泥状过滤取汤。将过滤的汤倒回锅内，加入热牛奶，继续小火加热，不可以煮沸，再加入搓碎的车达奶酪搅拌，直到奶酪全部融化，再加入奶油，用盐和白胡椒粉调味。

Blanch the broccoli, white mushrooms and green peas, keep some for garnish, and roughly chop the rest. Melt butter in the pan, add diced shallot, diced celery, white mushrooms, broccoli and green peas and saute before adding flour. Continue to stir until thick, then slowly add chicken broth, heat to a boil, keep stirring until thick, pour into a blender and blend into a puree, strain and keep the soup. Pour the strained soup back into the pot, add the hot milk, continue to simmer, do not boil, then add the grated cheddar cheese and stir until the cheese is completely melted. Finally, add the cream, seasoning it with salt and white pepper powder.

❸ 将车达奶酪西蓝花汤倒入盛汤的器皿内，从低温水槽中取出龙虾球一切为二放置于汤内，用西蓝花、青豆、白蘑菇装饰即成。

Pour the cheddar cheese broccoli soup into the soup bowl, take out the lobster balls from the sous vide and cut them in half, put on top of the soup. Use broccoli, green peas and white mushrooms as garnish.

技术解析 Technical resolution

❶ 圆葱粒、西芹粒、白蘑菇、西蓝花和青豆在锅中煸炒时，蔬菜不可以变成黄色，之后加面粉继续搅拌至黏稠时注意火候以免变色煳锅底。

Sweat the diced shallot, diced celery, white mushroom, broccoli and green peas in the pan without letting them yellow. After that, add flour and continue to cook until it becomes thick.

❷ 重新给车达奶酪西蓝花汤加温时，切不可以煮沸，否则奶酪会结块或者分开。

Carefully reheat the cheddar cheese broccoli soup, but do not let it boil, otherwise the cheese will curdle or separate.

二十九、龙虾尾肉奶油玉米汤
No.29 Lobster Tail with Cream Corn Soup

小记 Notes

加拿大新斯科舍省盛产龙虾，此菜正是使用加拿大盛产的龙虾和本土的甜玉米烹制而成。此菜用中餐鸡肉玉米汤和西式龙虾搭配而成。

The Canadian local lobster ingredients sourced from Nova Scotia. This dish is cooked by using Canadian local lobsters and sweet corn. This dish is created by using Chinese chicken corn soup pair with western-style cooked lobster.

主料 Main materials

龙虾尾，鲜奶油玉米粒。

lobster tail, fresh cream corn kernels.

配料 Ingredients

红灯笼椒，干紫菜，龙虾子。

red pepper, dried nori, lobster roe.

调料 Seasonings

圆葱粒，盐，白醋，糖，黑胡椒，白胡椒粉，鸡蛋清，鲜奶油，香芹末，柠檬，芥花油，鸡汤。

diced shallot, salt, white vinegar, sugar, black pepper, white pepper powder, egg white, fresh cream, minced parsley, lemon, canola oil, chicken broth.

制作过程 Methods

1. 将龙虾尾去壳，将皮和肉分离，取一只龙虾肉然后撒上盐和黑胡椒，用保鲜膜卷成圆形，放入冷藏室备用。

   Remove the shell from the lobster tail and separate the skin and meat. Take a piece of lobster meat and sprinkle it with salt and black pepper. Roll it into a circle with saran wrap and put it in the fridge for later use.

2. 将余下的龙虾肉加入鸡蛋清、鲜奶油、盐、柠檬汁，用搅拌机打成龙虾泥，然后放入柠檬皮末和香芹末轻轻搅拌均匀，装进裱花袋中备用。

   Add the remaining lobster meat, egg white, fresh cream, salt and lemon juice into a food processor and grind it into lobster paste. Fold minced lemon zest and minced parsley into the paste. Put it into a piping bag for later use.

3. 将红灯笼椒洗净片成薄片，用白醋和糖略腌。

   Wash the red pepper and slice it thinly. Marinate it with white vinegar and sugar.

❹ 将龙虾皮铺在眼睛形状的模具中，然后挤入龙虾泥，铺上红灯笼椒片，再挤入龙虾泥，铺上干紫菜片，放上卷好的龙虾肉卷，两侧挤入龙虾泥盖上龙虾皮，封好模子，用保鲜膜裹紧，放入50℃的蒸箱中蒸40分钟取出。

Spread the lobster skin in an eye shape mold and place the lobster paste on top of the lobster skin. Place sliced red peppers on top of the paste, followed by another layer of lobster paste, then the nori and the lobster roll on top of it. Place the lobster paste on both sides of the lobster meat roll and cover it with lobster skin. Wrap the mold tightly with saran wrap. Steam in a steamer at 50°C for 40 minutes and take it out.

❺ 锅内放入芥花油，煸香圆葱粒，放入鲜奶油玉米粒、鸡汤烧开，加入盐、白胡椒粉调味，倒入搅拌机内搅拌至均匀无颗粒状。

Put canola oil in a pot, add diced shallot to sauté until fragrant, then add fresh cream corn kernels and chicken broth. Bring it to a boil then add salt and white pepper powder to taste. Pour it into a blender to blend until smooth without lumps.

❻ 装盘：将眼睛形状的龙虾肉切成1厘米厚的片，平放到盘中，然后将玉米汤均匀地浇在四周，撒上龙虾子即可。

Plate: Slice the lobster into 1 cm thickness and place them flat on a shallow soup bowl. Then pour the corn soup evenly around it and sprinkle it with lobster roe.

技术解析 Technical resolution

龙虾肉要用盐和黑胡椒腌制，然后用保鲜膜卷成圆形，放入冷藏室备用。

Marinate the lobster with salt and black pepper, then roll it in a circle with saran wrap and put it in the freezer.

三十、中式白鱼汤
No.30 Chinese-style Fish Consommé with Whitefish

小记 Notes

在加拿大西餐中用海产品做菜肴非常普遍，但是很少会使用淡水鱼做菜。加拿大湖泊中盛产白鱼。此菜是运用中餐的酸辣淡水鱼汤味型搭配低温慢煮淡水白鱼的一个开胃前菜，咸鲜酸辣，鱼肉滑嫩。

In Canadian western cuisine, it is very common to use seafood as ingredients, but freshwater fish is rarely used. Whitefish is abundant in Canadian lakes. This dish uses Chinese-style sour and spicy freshwater fish soup flavor to match with slowly cook at sous vide freshwater whitefish as an appetizer. Salty, umami, sour and spicy, and the fish meat is tender and smooth.

主料 Main materials

淡水白鱼。

freshwater whitefish.

配料 Ingredients

芦笋。

asparagus.

调料 Seasonings

盐，酱油，白醋，白胡椒粉，清鱼汤，圆葱粉，大蒜粉，烟熏辣椒粉，粘肉粉，鱼松，黑胡椒。

salt, soy sauce, white vinegar, white pepper powder, clear fish soup, shallot powder, garlic powder, smoked chili powder, meat glue powder, fish floss, black pepper.

制作过程 Methods

1. 将白鱼去骨去皮洗净切成条状，在刀口处均匀地撒上盐、粘肉粉、大蒜粉、圆葱粉、烟熏辣椒粉和黑胡椒的混合物，再将刀口处黏合在一起，用保鲜膜卷紧成自然形状，放入50℃的恒温水槽中40分钟，捞出切成1.5厘米厚的片备用。

   Remove the bones and skin of the whitefish and wash. Cut it into strips and sprinkle salt, meat glue powder, garlic powder, shallot powder, smoked chili powder and black pepper evenly on the knife edge. Then put back the knife edge together and wrap it tightly into a natural shape with saran wrap. Put it in a 50℃ sous vide for 40 mins. Take it out and slice it into 1.5 cm thick for later use.

❷ 将芦笋洗净后焯水，再纵向切开备用。

Wash and blanch the asparagus and cut it lengthwise for later use.

❸ 锅内倒入清鱼汤，烧开加入少许的酱油、盐、白胡椒粉和白醋调味。

Pour clear fish soup into the pot, boil it and season it with a little soy sauce, salt, white pepper powder and white vinegar.

❹ 装盘：将芦笋放入浅汤盘中，取两片鱼肉放入汤盘中间，将调好味的鱼汤从鱼肉的四周倒入，在鱼肉上面撒上鱼松即可。

Plate: Put the asparagus in a shallow soup plate, take two slices of fish meat and put them in the middle of the soup plate. Pour the well-seasoned fish soup from around the fish meat and sprinkle with fish floss on top of the fish meat.

技术解析 Technical resolution

白鱼肉在刀口处用盐、粘肉粉、大蒜粉、圆葱粉、烟熏辣椒粉和黑胡椒的混合物黏合在一起，放入恒温水槽中静置40分钟。

The whitefish meat is put together with a mixture of salt, meat glue powder, garlic powder, shallot powder, smoked chilli powder and black pepper at the edge of the knife, and left in a constant temperature sink for 40 minutes.

Part

第二部分

杂碎菜

Chop Suey Dishes

“杂碎菜”释义

“杂碎”一词，在《现代汉语词典》（第7版）中的解释是“煮熟切碎供食用的牛、羊等的内脏”。

在19世纪末到20世纪的大部分时间里，美国、加拿大的中餐馆被称为“杂碎”餐馆，但近几十年来，随着正宗的中餐馆的不断出现与增多，“杂碎”餐馆已经大不如前，但仍然可以见到挂着“CHOP SUEY”（杂碎）字样的中餐馆。而且在厨师的传统认知中，也把中餐里的一部分菜肴称为“杂碎菜”。

关于“杂碎”一词，据国内厨师界传说，是清代的李鸿章在1896年出访美国时，在招待美国客人的宴席上，有一道用多种原料融合一起（即中国的“什锦菜”）烹饪加工的菜肴，但美国客人询问其菜肴名称的时候，由于李鸿章不知道菜肴的名字，当即随口称之为“杂碎菜”。从此，在美国、加拿大就有了中餐馆经营“杂碎菜”的出现。于是，“杂碎菜”便成为中餐馆的代名词或者是中国菜的代名词。

但据有关学者考证，“杂碎”一词不见于晚清以来的中国辞书中，或许是“外来语”。

据学者研究，“Chop suey”来自粤语“杂碎”，起源于19世纪末的美国。1903年梁启超游历美国，在《新大陆游记》一文中对“杂碎馆”的起源有详细的记述，应当是第一手资料。原文如下：

杂碎馆自李合肥游美后发生。此前西人足迹不履唐人埠，自合肥至后一到游历，此后来者如鲫。西人好奇家欲知中国人生活之程度，未能至亚洲，则必到纽约唐人埠一观焉。合肥在美思中国饮食，属唐人埠之酒食店进馔数次。西人问其名，华人难于具对，统名之曰“杂碎”，自此杂碎之名大噪。仅纽约一隅，杂碎馆三四百家，遍于全市。此外东方各埠，如费尔特费、波士顿、华盛顿、芝加高、必珠卜诸埠称是。全美国华人衣食于是者凡三千余人，每岁此业所入可数百万，蔚为大国矣。

中国食品本美，而偶以合肥之名噪之，故举国嗜此若狂。凡杂碎馆之食单，莫不大书“李鸿章杂碎”“李鸿章面”“李鸿章饭”等名。因西人崇拜英雄性及好奇性，遂产出此物。李鸿章功德之在粤民者，当惟此为最矣，然其所谓杂碎者，烹饪殊劣，中国人从无就食者。[1]

梁启超在书中也记录了李鸿章出访美国，把中餐的“杂碎菜”推向了一个高超。而且据梁启超的统计表明，当时仅在美国的纽约就有三四百家“杂碎馆”，几乎遍布全市。全美华人以“杂碎馆”为生的有3000多人，书中有“每岁此业收入可数百万”，足见“杂碎馆”之普及。

但梁启超在书中还说：“然其所谓杂碎者，烹饪殊劣，中国人从无就食者”。由此可知，“杂碎”不是正宗的中国烹饪，由此可以证明20世纪初中国本土不存在“杂碎”一词。“杂碎”一词很有可能就是通过梁启超的《新大陆游记》肇始传入中国的。

---

[1] 梁启超．新大陆游记［M］．北京：中国文史出版社，2020：45.

还有一点需要说明的是，有一张现存美国最早的“杂碎”餐馆的菜单，是1879年波士顿“宏发楼”的菜单，现存纽约美洲华人博物馆，说明“杂碎”一词在李鸿章访美之前已经存在，后来因为李鸿章访美而风行全美。

一、四川牛肉
No.1 Fried Ginger Beef

小记 Notes

此菜是加拿大中西餐结合菜中最受加拿大本地普通消费者喜爱的菜品之一，是以川菜生姜牛肉丝为基础做成的。

This dish is one of the most popular dishes among ordinary Canadian consumers that combines Chinese and western cuisine in Canada. This dish is based on the Sichuan dish shredded beef with ginger.

主料
Main materials

牛肉条。
beef strips.

配料
Ingredients

圆葱丝，青红辣椒丝。
shredded shallot, shredded green and red peppers.

调料
Seasonings

芥花油，淀粉，面粉，鸡蛋，啤酒，姜汁，蒜蓉辣椒酱，蔗糖，香醋，水，盐。
canola oil, starch, flour, eggs, beer, ginger juice, garlic chili sauce, cane sugar, red vinegar, water, salt.

制作过程 Methods

❶ 将牛肉条放入流动的冷水中冲净血水，捞出控干水分，加入啤酒、姜汁腌制2小时，加入鸡蛋液搅匀，均匀地裹上面粉和淀粉，放入油锅中炸至浅黄捞出，然后复炸至金黄色捞出控油。

Put the beef strips into flowing cold water to rinse off the blood, pull the beef out of the water and dry. Marinate with beer and ginger juice for two hours, then add egg liquid and mix well. Coat with flour and starch, fry in oil until light yellow then refry until golden yellow and take out.

❷ 炒锅中放入少许芥花油，放入圆葱丝煸炒，加入姜汁、蒜蓉辣椒酱、盐、蔗糖、香醋、水熬至略微黏稠，放入青红辣椒丝，再放入炸好的牛肉条翻炒，将酱汁均匀裹在牛肉条上。

Put the canola oil in the wok and stir-fry the shredded shallot. Add ginger juice, garlic chili sauce, salt, cane sugar, red vinegar, water until slightly thick, add shredded green and red peppers and then add fried beef strips to stir-fry, wrap the sauce evenly on the beef strips.

❸ 装盘，放上青红辣椒细丝点缀。

Plate and sprinkle with shredded green and red peppers.

技术解析 Technical resolution

此菜的关键点在于腌制牛肉一定要用啤酒和姜汁，不可以用嫩肉粉或食粉，加拿大本土人吃牛肉要保持牛肉的原味。再者就是上桌后牛肉条需长时间保持酥脆的口感，所以在炸制过程中要注意炸的时间。

The key point of this dish, pickled beef must use beer and ginger juice, not tender meat powder or food powder, native Canadians eat beef to maintain the original flavor of beef. Moreover, the crispy texture of beef strips will last for a long time after serving, so pay attention to the time of frying.

牛肉条炸后的效果如右图所示。

The effect of frying beef strips is shown in the right figure.

炸后的牛肉条
Fried beef strips

二、蜜桃虾
No.2 Peaches Shrimp with Mayo Sauce

小记 Notes

此菜是加拿大中西餐结合菜中最受加拿大本地普通消费者喜爱的菜品之一，是在中餐酥炸虾仁的基础上改良的，加入加拿大本地人日常生活中离不开的美乃滋，使此菜的口感具有油而不腻、百吃不厌的特点。

This dish is one of the most popular dishes that combines Chinese and western cuisine in Canada. It is an improved version of Chinese fried shrimp, which adds mayonnaise, a staple in the daily life of Canadians. Make the taste of this dish oily but not greasy, and never tire of eating it.

主料 Main materials

老虎虾。

tiger shrimp.

配料 Ingredients

罐装黄蜜桃。

canned yellow peaches.

调料 Seasonings

芥花油，淀粉，鸡蛋清，盐，美乃滋，雪碧。

canola oil, starch, egg white, salt, mayonnaise, sprite.

制作过程 Methods

❶ 将老虎虾去头，去皮，去沙洗净，加入盐、鸡蛋清拌匀，然后加入淀粉，使每个虾仁均匀地裹上淀粉，放入油锅中炸至浅黄捞出，然后复炸至皮脆，捞出控油。

Remove the head, shell and sand from the tiger shrimp, add salt, egg white and mix well, then add starch to evenly coat each shrimp. Fry in oil until light yellow, take out and fry again until crispy.

❷ 调拌盆中放入美乃滋、雪碧、黄蜜桃搅拌均匀，倒入炸好的虾仁调拌均匀，装盘。

Put mayonnaise, sprite and canned yellow peaches in a mixing bowl and stir well. Pour in the fried shrimp and mix well. Serve on a plate.

技术解析 Technical resolution

在美乃滋加入雪碧、黄蜜桃之后，搅拌时注意美乃滋不要松澥。同时要注意放入的炸好虾仁的温度不可以过高，否则美乃滋将融化。

The key to this dish is to add sprite and yellow peach to the mayonnaise, and when stirring, make sure that mayonnaise doesn't loosen and that the fried shrimp is not too hot, or the mayonnaise will melt.

三、酥炸柠檬鸡
No.3 Crispy Fried Lemon Chicken

小记 Notes

柠檬鸡是使用中式传统的油炸方法，融合西式烹调中常用的柠檬、白兰地酒创作而成，是中西融合菜中鸡肉类的典型代表菜。

Lemon chicken is fried in a traditional Chinese frying method, fused with lemon and brandy wine, which is commonly used in western cooking. It is a typical representative dish of the chicken category in Chinese and western fusion cuisine.

主料 Main materials

鸡胸脯肉。

chicken breast.

配料 Ingredients

红薯粉，柠檬片。

sweet potato vermicelli, lemon slices.

调料 Seasonings

芥花油，黄油，葱姜汁，盐，白兰地，白糖，柠檬汁，柠檬皮，橙汁，淀粉，鸡蛋，面粉，泡打粉。

canola oil, butter, onion and ginger juice, salt, brandy, white sugar, lemon juice, lemon zest, orange juice, starch, eggs, flour, baking powder.

制作过程 Methods

❶ 将鸡胸脯肉去筋膜，加入葱姜汁、盐、白兰地、柠檬皮拌匀，放入冷藏室腌制0.5小时备用。

Remove the fascia of chicken breast, add onion and ginger juice, salt, brandy and lemon zest and mix well, marinate in the freezer for 0.5 hours.

❷ 将面粉、淀粉、鸡蛋、泡打粉、水、芥花油调成糊状，放入冷藏室备用，随用随取。

Mix flour, starch, eggs, baking powder, water and canola oil into a batter and put it in the cooler for reserve, and take whatever you want.

❸ 锅内倒入油加热到七成热，将鸡肉拍匀面粉，然后裹匀面糊，放入油锅炸至金黄，外酥里嫩。

Heat the oil in a pan until 70% hot, pat chicken with flour, then coat with batter and fry until golden brown, crispy outside and tender inside.

❹ 净锅放入水、柠檬汁、橙汁、白糖烧开，加入淀粉勾芡，放入黄油搅拌均匀成柠檬酱备用。

Put water, lemon juice, orange juice, white sugar into a clean pot, bring to a boil to thicken the starch. Add butter and mix well to make lemon sauce. Set aside.

❺ 将红薯粉炸至酥脆，垫入盘底。放上炸好的鸡肉，淋上柠檬酱，用柠檬片点缀。

Deep fried sweet potato vermicelli until crispy and place on the bottom of the plate. Top with fried chicken breasts, drizzle with lemon sauce and garnish with lemon slices.

技术解析 Technical resolution

鸡胸脯肉一定要冷藏腌制0.5小时以上，鸡胸脯肉和红薯粉都要炸至酥脆。

chicken breast must be marinated for more than 0.5 hours in the fridge, chicken breast and sweet potato vermicelli are fried until crisp.

四、菠萝咕噜肉
No.4 Sweet and Sour Pork with Pineapple

小记 Notes

菠萝咕噜肉也就是我们中餐中的菠萝咕咾肉。在加拿大的中餐统称菠萝咕噜肉，是西方人特别喜欢吃的菜肴，进餐馆必点之菜。但是和传统的菠萝咕咾肉还是略有不同。

Sweet and sour pork with pineapple, also known as sweet and sour pork with pineapple in Chinese cuisine. In Canada, it is generally called sweet and sour pork with pineapple, and it is a dish that westerners particularly like to eat, and it is a must-order dish when dining in a restaurant. However, it is slightly different from traditional sweet and sour pork with pineapple.

主料 Main materials

猪里脊。

pork tenderloin.

配料 Ingredients

青红椒丝，圆葱丝，胡萝卜丝，罐装菠萝粒，鸡蛋。

shredded green and red pepper, shredded shallot, shredded carrots, canned diced pineapple, eggs.

调料 Seasonings

芥花油，面粉，淀粉，盐，生抽，胡椒粉，白糖，黑醋，番茄酱。

canola oil, flour, starch, salt, light soy sauce, pepper powder, white sugar, black vinegar, tomato sauce.

制作过程 Methods

❶ 将猪里脊切成块，加入盐、生抽、胡椒粉腌制，放入蛋液搅拌均匀，裹匀面粉和淀粉的混合粉。

Cut the pork tenderloin into pieces and marinate with salt, light soy sauce and pepper powder. Add egg liquid and stir well, wrap the flour and starch into a batter.

❷ 锅中加入芥花油烧至七成热，放入裹好的猪肉炸至金黄，捞出控油，装盘。

Add canola oil in the pot to cook 70% heat, deep-fry the wrapped pork until golden, remove the oil and plate.

❸ 锅中加入水、白糖、黑醋、番茄酱、盐烧开之后，加入水淀粉勾浓芡，放入青红椒丝、圆葱丝、胡萝卜丝、罐装菠萝粒拌匀，淋上明油。浇在炸好的猪里脊上即可。

Add water, white sugar, black vinegar, tomato sauce and salt to the wok. After boiling, thicken the sauce with water starch. Add shredded green and red pepper, shredded shallot, shredded carrots, canned diced pineapple and pour a little oil in and mix well. Then pour the sauce on top of the fried pork tenderloin.

技术解析 Technical resolution

腌制要入味，猪肉要炸至金黄色熟透。

Marinate thoroughly, deep-fry pork until golden brown and cooked fully.

五、酥炸鸡蛋卷
No.5 Crispy Fried Egg Roll

小记 Notes

酥炸鸡蛋卷是借鉴中餐的炸春卷改变而来，用发酵好的面粉做成皮包裹熟馅料来代替传统的炸春卷，是西方人喜爱的开胃前菜。

Crispy fried egg roll is derived from Chinese fried spring roll, using fermented flour to make the skin to wrap cooked fillings instead of traditional fried spring roll. It is a popular appetizer among westerners.

主料 Main materials

猪肉碎。

minced pork.

配料 Ingredients

胡萝卜，圆葱，番茄，卷心菜。

carrots, shallot, tomatoes, cabbage.

调料 Seasonings

芥花油，葱花，姜末，料酒，酱油，盐，味精，面粉，鸡蛋，酵母，酸甜柠檬酱。

canola oil, chopped green onion, minced ginger, cooking wine, soy sauce, salt, MSG, flour, eggs, yeast, sweet and sour lemon sauce.

制作过程 Methods

❶ 将面粉、酵母、盐放入盆中拌匀，加入鸡蛋和温水搅拌均匀，放在室温发酵。将番茄切成小块，圆葱、胡萝卜、卷心菜切成丝。

Mix flour, yeast and salt in a bowl. Add eggs and warm water and mix well. Leave at room temperature to ferment. Cut tomatoes into small pieces. Cut shallot, carrots and cabbage into shreds.

❷ 净锅倒入芥花油加热，先放入猪肉碎，再放入葱花、姜末和番茄煸炒，最后加入圆葱丝、胡萝卜丝、卷心菜丝继续煸炒至熟，用料酒、酱油、盐、味精调味翻炒均匀，倒入事先准备好的盆里凉凉。

In a wok, heat canola oil and add minced pork first, then add chopped green onion, minced ginger and tomatoes to stir-fry. Lastly add shredded shallot, shredded carrots and shredded cabbage to continue stir-frying until cooked. Season with cooking wine, soy sauce, salt and MSG. Stir-fry evenly and pour into a prepared bowl to cool.

❸ 取出发酵好的面团搅拌均匀，做成需要大小的剂子，然后制作成薄皮，卷入凉凉的馅料，裹匀鸡蛋液，放入热油锅中炸至金黄色捞出，装盘，配上酸甜柠檬酱。

Take out the fermented dough and mix well. Make the required size of dough and pull it into thin skins. Roll in the cooled filling. Coat with egg liquid and fry in hot oil until golden brown. Plate and serve with sweet and sour lemon sauce.

技术解析 Technical resolution

发酵糊要调匀，温室发酵；鸡蛋卷炸至深金黄色酥脆。

The batter should be mixed well and fermented in the room temperature; the egg roll should be deep-fried until golden brown and crispy.

六、炒牛肉西蓝花
No.6 Fried Beef with Broccoli

小记 Notes

炒牛肉西蓝花是北美人非常喜欢的一道中式杂碎菜。牛肉是北美人最喜欢的肉类，西蓝花则是北美人最喜爱的蔬菜，可以熟食也可以生吃。将两种原料结合，运用中餐的烹饪技巧进行烹制，在北美中西融合中餐馆中很常见。

Fried beef with broccoli is a favorite dish of Chinese chop suey in North Americans. Beef is the favorite meat of North Americans, and broccoli is their favorite vegetable, which can be eaten cooked or raw. A combination of the two ingredients using Chinese cooking techniques, common in North American fusion Chinese restaurants.

主料 Main materials

牛里脊肉，西蓝花。

beef tenderloin, broccoli.

配料 Ingredients

胡萝卜。

carrots.

调料 Seasonings

芥花油，葱花，蒜末，料酒，酱油，盐，味精，小苏打，水淀粉。

canola oil, chopped green onion, minced garlic, cooking wine, soy sauce, salt, MSG, baking soda, water starch.

制作过程 Methods

❶ 将牛里脊肉切成薄片，用流动水冲去血水，捞出控干水分，放入盆中用料酒、小苏打和盐腌制。

Cut the beef tenderloin into thin slices. Rinse off the blood with running water and dry. Marinate with cooking wine, baking soda and salt in a bowl.

❷ 将西蓝花切成小块，胡萝卜去皮切成片。

Cut broccoli into small pieces. Peel and slice carrots.

❸ 锅内放水烧开，将牛肉和西蓝花、胡萝卜先后焯水。净锅倒入芥花油加热，爆香葱花和蒜末，放入酱油和焯好水的牛肉、西蓝花、胡萝卜，用盐、味精调味翻炒均匀，用水淀粉勾芡即可。

Boil water in a pot and blanch beef, broccoli and carrots in turn. Heat canola oil in a wok, sweat chopped green onion and minced garlic until fragrant, then add soy sauce and blanched beef, broccoli and carrots. Season with salt and MSG. Stir-fry evenly and thicken with water starch.

技术解析 Technical resolution

牛肉要用流动的清水冲去血污，并在炒前焯水，以保持牛肉原有的清香。

The beef should be washed with running water to remove the blood stains, and blanch before frying to keep the original smell of beef.

七、酸辣汤
No.7 Hot and Sour Soup

小记 Notes

酸辣汤是中西融合汤类最具代表性的菜品之一。此汤是传统粤菜酸辣汤的延伸，加入了虾仁、叉烧和鸡肉。此汤的口味既适合中国人，又满足西方人的味蕾需求。

Hot and sour soup is one of the most representative fusion soups that combines Chinese and western cuisine. This soup is an extension of traditional Cantonese hot and sour soup, with the addition of shrimp, char siu and chicken. The taste of this soup is suitable for both Chinese people and westerner.

主料 Main materials

虾仁，猪叉烧肉，鸡肉。

shrimp, char siu, chicken.

配料 Ingredients

豆腐，木耳，竹笋，胡萝卜，青豆，青葱粒。

tofu, black fungus, bamboo shoots, carrots, green peas, diced green onion.

调料 Seasonings

芥花油，鸡汤，辣椒酱，酱油，老醋，淀粉，胡椒粉，盐，鸡粉，香油。

canola oil, chicken broth, chili sauce, soy sauce, vinegar, starch, pepper powder, salt, chicken powder, sesame oil.

制作过程 Methods

❶ 将虾仁洗净，猪叉烧肉切成丝，鸡肉切成丝，用水和淀粉上浆备用。

Wash and clean the shrimp. Cut the char siu and chicken into shreds. Size with water and starch. Set aside.

❷ 将豆腐切成细长条，木耳、竹笋、胡萝卜切丝，然后和青豆一起用沸水焯水。

Cut the tofu into thin strips. Shred the black fungus, bamboo shoots and carrots. Blanch them together with green peas.

❸ 净锅放入少量芥花油加热，放入辣椒酱炒香，加入鸡汤烧开，放入豆腐条、木耳丝、竹笋丝、胡萝卜丝、青豆，用酱油、老醋、盐、鸡粉、胡椒粉调味，加入淀粉勾芡，淋入香油，出锅撒上青葱粒即可。

Heat a small amount of canola oil in a clean pot and stir-fry chili sauce until fragrant. Add chicken broth and bring to a boil. Add tofu strips, black fungus shreds, bamboo shoot shreds, carrots shreds, green peas, add soy sauce, vinegar, salt, chicken powder and pepper powder to season. Thicken with starch and drizzle with sesame oil before serving. Sprinkle with diced green onion.

技术解析 Technical resolution

用淀粉勾芡时，为避免淀粉在汤中结块，两只手需要同时运用。一只手搅动锅内的汤，另一只手慢慢地将淀粉倒入锅内。

When using starch to thicken the soup, both hands need to be used simultaneously to avoid lumps forming in the soup. One hand pushes the soup in the pot while the other hand slowly pours in the starch.

八、炸鸡球
No.8 Deep Fried Chicken Ball

小记 Notes

炸鸡球也是中西融合杂碎菜中代表性的菜品之一。西方人比较喜爱炸制食品，也比较喜欢发酵的食品，所以此菜将两者结合，用发酵面糊代替普通的炸面糊，再配以柠檬酱汁或者糖醋酱汁。

Deep fried chicken ball is also one of the typical representative dishes in Chinese and western fusion chop suey dish. Westerners prefer fried foods and fermented foods, so this dish combines the two, using fermented batter instead of ordinary fried batter, and is served with lemon sauce or sweet and sour sauce.

主料 Main materials

鸡腿肉。

chicken thigh.

配料 Ingredients

面粉。

flour.

调料 Seasonings

芥花油，橄榄油，盐，黑胡椒粉，酵母，淀粉，泡打粉，糖，柠檬酱汁（或糖醋酱汁）。

canola oil, olive oil, salt, black pepper powder, yeast, starch, baking powder, sugar, lemon sauce (or sweet and sour sauce).

制作过程 Methods

1. 将鸡腿肉去皮，去筋膜，切成大拇指盖大小的块状，用盐、黑胡椒粉腌制备用。
   Remove the skin and fascia from the chicken thigh and cut into thumb-sized pieces. Marinate with salt and black pepper powder.
2. 将面粉、淀粉、泡打粉、糖、酵母、温水、橄榄油混合均匀成面糊，放在室温下30分钟，使面糊发酵。
   Mix flour, starch, baking powder, sugar, yeast, warm water and olive oil to make a batter. Leave at room temperature for 30 minutes to ferment the batter.
3. 锅内倒入芥花油加热至七成时，将鸡肉拍匀面粉，然后裹匀面糊，放入油锅炸至金黄，外酥里嫩即可。
   Heat canola oil in a wok to 70% hot. Coat the chicken with flour and then coat it with batter. Fry in oil until golden brown and crispy on the outside and tender on the inside.

❹ 上桌时配以柠檬酱汁或糖醋酱汁即可。

Serve with lemon sauce or sweet and sour sauce.

技术解析 Technical resolution

鸡球要制作的大小均匀，色泽为深金黄色。

Chicken ball should be uniform in size and dark golden in color.

酱汁制作
Making sauces

（1）柠檬酱汁：净锅放入水、柠檬汁、橙汁、糖烧开，用淀粉勾芡，加入黄油搅拌均匀。

Lemon sauce: In a pot, add water, lemon juice, orange juice and sugar. Boil and thicken with starch. Add butter and mix well.

（2）糖醋酱汁：净锅放入白糖、白醋、番茄酱，小火熬至黏稠。

Sweet and sour sauce: In a pot, add white sugar, white vinegar and tomato paste and bring to boil. Then simmer over low heat until thick.

九、蒙特利尔花生酱饺子
No.9 Montreal Peanut Butter Dumpling

小记 Notes

花生酱饺子（馄饨）在加拿大东部的蒙特利尔是非常流行的。此菜品由花生酱、蜂蜜和酱油混合的酱汁浇在饺子或者馄饨上制作而成，借鉴了中餐的红油抄手，更符合蒙特利尔人的口味。

Peanut butter dumplings (wontons) are very popular in Montreal in eastern Canada. The dish consisting of wonton or dumpling covered with peanut butter, honey and soy-sauce based sauce, most likely based-on the Sichuan dish hongyou chaoshou (folded-dumplings in red-oil) and reinterpreted for the tastes of Montrealer.

主料 Main materials

猪肉馅。

minced pork.

配料 Ingredients

饺子皮。

dumpling wrappers.

调料 Seasonings

芥花油，黄油，葱花，葱末，姜末，蒜末，蜂蜜，酱油，鱼露，白醋，盐，味精，花生酱，香油，辣椒油，熟芝麻。

canola oil, butter, chopped green onion, minced green onion, minced ginger, minced garlic, honey, soy sauce, fish sauce, white vinegar, salt, MSG, peanut butter, sesame oil, chili oil, roasted sesame.

制作过程 Methods

1. 不锈钢盆内放入猪肉馅，加入葱末、姜末、酱油、鱼露、盐、味精、香油，调和均匀成馅料。用饺子皮包裹馅料做成饺子。放入开水中煮熟捞出装盘。

   In a bowl, mix minced pork, minced green onion, minced ginger, soy sauce, fish sauce, salt, MSG and sesame oil to make the filling. Wrap the filling in the dumpling wrappers and make the finished dumpling. Put them in boiling water and cook until done.

❷ 锅内放入黄油，加入姜末、蒜末，用中火煸香，加入少许水，放入蜂蜜、酱油、鱼露、花生酱、白醋、香油、辣椒油混合均匀。浇在饺子上，撒上熟芝麻和葱花。

Add butter to the pot. Stir-fry minced ginger and garlic over medium heat until fragrant. Add water and honey, soy sauce, fish sauce, peanut butter, white vinegar, sesame oil and chili oil and mix well. Pour it on the dumplings and sprinkle with rosted sesame and chopped green onion.

技术解析 Technical resolution

肉馅要调制均匀，饺子要大小均匀，形态一致。

The minced meat should be mixed and beaten evenly, and the dumplings should be made of the same size and shape.

十、鸡肉蛋芙蓉
No.10 Chicken Egg Foo Young

小记 Notes

鸡肉蛋芙蓉在加拿大和美国的杂碎餐馆中常见。使用中式鸡蛋卷搭配一些蔬菜再配以西式的肉汁而成，也可以根据顾客要求加入鸡肉、牛肉、猪肉或者海鲜制作成肉类芙蓉或者海鲜芙蓉。制作芙蓉菜肴时烹调方法可以选择铁板煎或者油炸。

Chicken egg foo young is commonly found in Chinese and western fusion restaurants in Canada and the United States. It uses Chinese-style egg rolls with some vegetables and western-style gravy. It can also be made with chicken, beef, pork or seafood to make meat foo young or seafood foo young according to customer requirements. The cooking method for making foo young dishes can be either pan frying or deep frying.

主料
Main materials

鸡胸肉。

chicken breast.

配料
Ingredients

绿豆芽。

mung bean sprouts.

调料
Seasonings

芥花油，黄油，葱花，圆葱粒，蒜末，老抽，盐，味精，糖，白胡椒粉，面粉，鸡蛋，香油，鸡汤。

canola oil, butter, chopped green onion, diced shallot, minced garlic, dark soy sauce, salt, MSG, sugar, white pepper powder, flour, eggs, sesame oil, chicken broth.

制作过程 Methods

1. 将鸡胸肉用开水煮熟捞出，切成小粒。绿豆芽粗切一下焯水，捞出挤干水分。

   Boil the chicken breast in hot water until cooked and cut into small pieces. Blanch the mung bean sprouts roughly and squeeze out the water.

2. 不锈钢盆内放入鸡肉、绿豆芽、葱花、盐、糖、味精、面粉、鸡蛋和香油，慢慢地混合均匀成芙蓉。

   In a bowl, mix chicken, mung bean sprouts, chopped green onion, salt, sugar, MSG, flour, eggs and sesame oil slowly mix well into foo young.

3. 锅烧热抹一层油，将芙蓉慢慢倒入，用盖子盖住。煎至两面金黄，内部熟透即可。

   Heat the pan and apply a layer of oil. Slowly pour the foo young onto the pan and cover with a lid. Fry until both sides are golden brown and cooked through.

4. 锅内放入黄油，加入圆葱粒、蒜末，中火煸至圆葱微黄，放入面粉继续搅拌，当面粉渐渐变成金黄色时，加入鸡汤、老抽、盐、味精、糖、白胡椒粉和香油搅拌均匀成略稀的面糊状，倒入装有密漏的盛器内过滤掉杂质，浇在鸡肉饼上即可。

   Add butter to the pan. Sweat diced shallot and minced garlic over medium heat until shallot is slightly yellow. Add flour and continue stirring. When the flour gradually turns golden brown, add chicken broth, dark soy sauce, salt, MSG, sugar, white pepper powder and sesame oil. Pour into a container with a leak and filter to remove impurities. Pour over the chicken patty.

技术解析 Technical resolution

调制的芙蓉泥料要稀稠均匀，用锅烙制时注意控制火候，不要上色过重。

The preparation of foo young paste to thin and even, with the pot sear ten pay attention to control the heat, do not color too heavy.

十一、纽芬兰虾炒面
No.11 Newfoundland Shrimp Chow Mein

小记 Notes

纽芬兰虾炒面实际上在整个菜中没有用到面条，是将卷心菜切成长条状代替了面条。此菜品在加拿大的纽芬兰与拉布拉多省的中餐馆中是比较流行的，是加拿大中西融合杂碎菜中的代表性菜品之一。

Newfoundland shrimp chow mein actually do not use noodles in the entire dish. Cabbage is cut into long strips to replace the noodles. It is quite popular in Chinese restaurants in Newfoundland and Labrador province in Canada. It is one of the representatives of Canadian Chinese and western fusion cuisine.

主料 Main materials

虾仁。

shrimp.

配料 Ingredients

卷心菜，胡萝卜，西蓝花，芹菜。

cabbage, carrots, broccoli, celery.

调料 Seasonings

芥花油，葱花，姜末，蚝油，酱油，蜂蜜，盐，味精，香油，水淀粉。

canola oil, chopped green onion, minced ginger, oyster sauce, soy sauce, honey, salt, MSG, sesame oil, water starch.

制作过程 Methods

1. 将虾仁洗净去其虾线备用。卷心菜切成长条状，胡萝卜斜刀切成薄片，芹菜切段，西蓝花掰成小块。锅内加水烧开放入盐，然后将虾仁、卷心菜、胡萝卜、芹菜、西蓝花放在一起焯水。

   Clean the shrimp and remove the shrimp sand line. Cut cabbage into long strips. Slice carrots diagonally into thin slices. Cut celery into segments. Break broccoli into small pieces. Boil water in a pot and add salt. Blanch shrimp, cabbage, carrots, celery and broccoli together.

2. 锅内放入芥花油，加入葱花、姜末中火煸香，加入蚝油、酱油，加入一点水，再放蜂蜜，放入焯水后的虾和蔬菜翻炒，用盐、味精调味，加入水淀粉勾芡，翻炒均匀之后淋入香油即可。

   Add canola oil in the pot, add chopped green onion and minced ginger and stir fry, add oyster sauce, soy sauce, add a little water, then add honey, then add shrimps and vegetables after boiling, season with salt and MSG, thicken with water starch, stir-fry evenly and pour in sesame oil.

技术解析 Technical resolution

虾仁要去净虾线，卷心菜焯水要保持翠绿。

The shrimps should be cleaned and the cabbage should be blanched and keep green color.

十二、窝馄饨汤
No.12 Wor Wonton Soup

小记 Notes

窝馄饨汤是加拿大中西合璧汤菜中最具有代表性的菜品之一，馄饨汤在制作过程中和传统的中餐是一样的，只是在辅料上增加了一些北美人喜欢的蔬菜。

Wor wonton soup is one of the most representative dishes in Canada that combines Chinese and western cuisine. The production process of wonton soup is the same as traditional Chinese cuisine, but some vegetables that North Americans like are added.

主料 Main materials

五花肉馅，馄饨皮。

minced pork belly, wonton wrapper.

配料 Ingredients

虾仁，西蓝花，菜花，白蘑菇，油菜心，胡萝卜粒，青豆，木耳，青葱粒。

shrimp, broccoli, cauliflower, white mushrooms, baby bok choy, diced carrot, green peas, black fungus, diced green onion.

调料 Seasonings

姜末，葱末，花雕酒，酱油，盐，鸡粉，鸡汤，鸡蛋液，香油。

minced ginger, minced green onion, Huadiao cooking wine, soy sauce, salt, chicken powder, chicken broth, egg liquid, sesame oil.

制作过程 Methods

1. 碗中放入五花肉馅，加入葱末、姜末、花雕酒、酱油、盐、鸡粉调成馅。将猪肉馅放入馄饨皮中间，馄饨皮边缘抹匀鸡蛋液，馄饨皮对折，捏实边缘，包成馄饨。

   Put minced pork belly in a bowl and add, minced green onion, minced ginger, Huadiao cooking wine, soy sauce, salt and chicken powder to make the filling. Put the pork filling in the middle of the wonton wrapper. Spread egg liquid evenly on the edge of the wonton wrapper. Fold the wonton wrapper in half and pinch the edges to make it into a wonton shape.

❷ 锅内加入鸡汤烧开，放入馄饨煮至八成熟，放入虾仁、木耳、西蓝花、菜花、胡萝卜粒、白蘑菇、油菜心、青豆，加入盐、鸡粉、香油调味，撒入青葱粒。

Add the chicken broth, bring to a boil and cook the wonton until 80% done, add shrimp, black fungus, broccoli, cauliflower, diced carrot, white mushrooms, baby bok choy, green peas, season with salt, chicken powder, sesame oil, sprinkle with diced green onion.

技术解析 Technical resolution

肉馅调制要保持一定的水分，馄饨包制要均匀。

The meat filling should be prepared with a certain amount of moisture, and the wontons should be wrapped evenly.

十三、培根蛋炒饭

No.13 Bacon and Egg Fried Rice

小记 Notes

培根蛋炒饭是在中餐蛋炒饭的基础上加上西方人早餐不可缺少的培根改良而来。中式的蛋炒饭的香气，加上西式培根的点缀，既符合中餐蛋炒饭的要求，同时也捕获了西方人的味蕾。

Bacon and egg fried rice is an improvement on Chinese egg fried rice by adding bacon that is indispensable for western breakfast. The aroma of Chinese egg fried rice, combined with the unique taste of western bacon, not only meets the requirements of Chinese egg fried rice, but also captures the taste buds of westerners.

主料 Main materials

米饭。

cooked rice.

配料 Ingredients

培根，芦笋，蛋黄液，青葱末。

bacon, asparagus, egg yolk, minced green onion.

调料 Seasonings

芥花油，盐，味精，酱油。

canola oil, salt, MSG, soy sauce.

制作过程 Methods

❶ 将培根切成粒状，芦笋顶刀切成薄片备用。

Cut the bacon into small pieces and slice the asparagus into thin slices for later use.

❷ 锅中倒入芥花油加热，放入培根煸香，倒入蛋黄液，在蛋黄液还没有凝固时，加入米饭快速翻炒，使蛋黄液均匀包裹米饭粒，然后加入芦笋片翻炒，加盐、味精、酱油调味，最后撒上青葱末出锅装盘。

Heat canola oil in a pan and stir-fry the bacon until fragrant. Add egg yolk and quickly stir-fry with rice before the egg yolk solidifies. Make sure the egg yolk coats the rice grains. Then add asparagus slices and stir-fry. Add salt, MSG and soy sauce to season. Lastly, sprinkle with minced green onion and serve.

技术解析 Technical resolution

蛋黄液入锅时，一定要离火，否则蛋黄液会快速凝固成蛋黄块，无法裹在米粒上。

Put the egg yolk into the wok, must be off the heat, otherwise the egg yolk liquid will quickly solidify into egg yolk chunks and cannot be coated on the rice grains.

十四、铁板红酒烧汁牛柳
No.14 Beef Tenderloin with Red Wine Hot Pan

小记 Notes

铁板红酒烧汁牛柳是加拿大中西餐融合菜中典型的菜品之一。此菜品是在中餐铁板牛柳的基础上进行改良，加入西餐中经常使用的红葡萄酒做成红酒烧汁。此菜色泽红润，红酒醇香浓郁。

Beef tenderloin with red wine hot pan (Sizzling beef tenderloin with red wine sauce) is one of the typical dishes of Chinese and western fusion cuisine in Canada. It is a fusion based on Chinese sizzling beef tenderloin, adding red wine commonly used in western cuisine to make red wine sauce. This dish has a red color and a rich aroma of red wine.

主料 Main materials

牛里脊。

beef tenderloin.

配料 Ingredients

圆葱丝，青豆，胡萝卜丝。

shredded shallot, green peas, shredded carrots.

调料 Seasonings

芥花油，黄油，老抽，生抽，红酒，淀粉，鸡蛋清，盐，鸡粉，糖，姜粉。

canola oil, butter, dark soy sauce, light soy sauce, red wine, starch, egg white, salt, chicken powder, sugar, ginger powder.

制作过程 Methods

❶ 将牛里脊片成大片，用姜粉腌制之后，冷水冲洗，吸干水分，加入盐、老抽、鸡蛋清、淀粉拌匀，表面用芥花油封住，放入冷藏室保存。

Cut the beef tenderloin into large pieces horizontally, marinate with ginger powder, rinse with cold water, pat dry the beef, add salt, dark soy sauce, egg white and starch and mix well. Seal the surface with canola oil and store in the cooler.

❷ 锅内倒入芥花油烧至五成热时，将牛柳放入油锅中滑透，捞出控油。另起锅放入黄油和芥花油的混合油，倒入红酒，待酒精挥发之后，放入一点水、老抽、生抽、糖、鸡粉调味，放入圆葱丝、胡萝卜丝和青豆烧开，用淀粉勾成流芡，放入牛柳翻炒均匀，倒在烧热的铁板上即可。

Place the canola oil into the wok and heat the oil to 110°C. Place the beef tenderloin into the oil and cook through. Remove them from oil and drain. In another wok, add a mixture of butter and canola oil, pour in red wine. After the alcohol evaporates, add not much water, dark soy sauce, light soy sauce, sugar and chicken powder to season. Add shredded shallot, shredded carrots and green peas to stir-fry. Add starch to make a thinner sauce. Add beef tenderloin and stir-fry evenly, then pour it onto a hot pan.

技术解析 Technical resolution

❶ 滑牛柳的时候油温不宜过高，时间不宜过久，否则肉质会变老。

When placing the marinated beef tenderloin into the oil should not stay too long and the oil temperature should not be too high, otherwise the meat will become chewy.

❷ 芡汁要勾成流芡。

The sauce should be a thinner sauce.

十五、上海粗面
No.15 Shanghai Fried Noodles

小记 Notes

上海粗面可以说是北美人最喜欢的中西融合式家常面食，主要特点是将传统上海炒面的细面条换成了有嚼劲的粗面条，同时增加了北美人喜爱的酱油，使口味转变成酱香型。

Shanghai fried noodles can be said to be one of the most popular Chinese and western fusion home-style noodles among north Americans. Its main feature is to replace the thin noodles of traditional Shanghai fried noodles with chewy thick noodles, while increasing the soy sauce that North Americans love, making the taste change into a soy sauce flavor.

主料 Main materials

粗面条。

thick noodles.

配料 Ingredients

卷心菜，香菇，香菜。

cabbage, shiitake mushrooms, cilantro.

调料 Seasonings

芥花油，葱末，蒜末，盐，生抽，老抽，鸡粉。

canola oil, minced green onion , minced garlic, salt, light soy sauce, dark soy sauce, chicken powder.

制作过程 Methods

1. 净锅加水烧开，放入粗面条煮熟捞出备用。卷心菜切成丝，香菇切成片。
   Boil water in a pot, add thick noodles and cook until done. Remove and set aside. Shred the cabbage and slice shiitake mushrooms.
2. 起锅放入芥花油，加入葱末、蒜末煸香，加入生抽、卷心菜、香菇继续煸炒，然后放入粗面条，加入盐、鸡粉、老抽调味，翻炒均匀即可。
   Heat canola oil in a clean pan, add minced green onion and minced garlic to sweat until soft. Add light soy sauce, cabbage and shiitake mushrooms and continue to stir fry, then add cooked noodles, add salt, chicken powder and dark soy sauce to season and stir fry evenly.
3. 装盘，用香菜点缀。
   Serve on a plate and garnish with cilantro.

技术解析 Technical resolution

在炒锅里加入粗面条时要快速翻炒，以免粗面条粘在锅底。

When adding thick noodles to the wok, it should be quickly stir-fried to avoid sticking to the bottom of the wok.

十六、蜜汁蒜香排骨
No.16 Honey Garlic Pork Ribs

小记 Notes

蜜汁蒜香排骨在中西融合的杂碎菜中也是西方人家喻户晓、比较喜爱的菜品之一，除排骨之外，还有蜜汁蒜香鸡翅也是非常流行的菜品，其烹调方式都相同。

Honey garlic pork ribs is one of the more popular dishes among westerners in the fusion of Chinese and western fusion cuisine. In addition to ribs, honey garlic chicken wings is also a very popular dish, with the same cooking method.

主料 Main materials

猪排骨。

pork ribs.

配料 Ingredients

姜末，蒜末。

minced ginger, minced garlic.

调料 Seasonings

芥花油，黄油，酱油，蜂蜜，红糖，盐，黑胡椒，淀粉，面粉。

canola oil, butter, soy sauce, honey, brown sugar, salt, black pepper, starch, flour.

制作过程 Methods

❶ 将猪排骨切成小段清洗干净，用盐、黑胡椒、湿淀粉腌制，然后裹匀干面粉备用。另起锅倒入芥花油烧热，将猪排骨放入油锅炸至表面酥脆金黄，捞出控油。

Cut the pork ribs into small sections and clean them. Stir in salt, black pepper and wet starch to marinate. Fold in dry flour. Set aside. Heat the canola oil in a wok, fry the pork ribs in the wok until the surface is crisp and golden brown, remove and control the oil.

❷ 锅内加入黄油、姜末、蒜末煸炒出香，加入酱油、蜂蜜、红糖和少许水，小火熬至黏稠，加入炸好的排骨，翻炒均匀即可。

Add butter, minced ginger and minced garlic in the pan and stir-fry until fragrant, add soy sauce honey, brown sugar and a little water, simmer until thick, add fried ribs, stir evenly can be.

技术解析 Technical resolution

猪排骨放入油锅炸时，要控制好油温，以内部熟透表面酥脆金黄色即可。

When deep frying the pork ribs, control the oil temperature so that the inside is cooked fully and the surface is crispy and golden brown.

十七、中式奶酪焗龙虾
No.17 Chinese-style Cheese Lobster

小记 Notes

加拿大盛产龙虾，也是西方人比较喜欢吃的海鲜。中式奶酪焗龙虾是典型的中西融合菜代表，运用中餐的烹饪技法加入西方人生活中不可缺少的奶酪融合而成。

Canada is rich in lobsters and is also a seafood that westerners like to eat. Chinese-style cheese lobster is a typical representative of Chinese and western fusion cuisine. It uses Chinese cooking techniques and adds cheese that is indispensable in westerners' lives.

主料 Main materials

活龙虾。

live lobster.

配料 Ingredients

圆葱末，奶酪。

minced shallot, cheese.

调料 Seasonings

芥花油，黄油，鲜奶油，盐，白胡椒粉，淀粉，柠檬汁。

canola oil, butter, fresh cream, salt, white pepper powder, starch, lemon juice.

制作过程 Methods

❶ 将龙虾宰杀干净，剁成小块，在龙虾刀口处拍上淀粉。

Clean and chop the lobster into small pieces and coat evenly the lobster with dry starch.

❷ 锅中倒入芥花油烧至七成热，放入龙虾炸至金黄，捞出控油。

Add canola oil to the pot and heat it to 70% heat. Deep fried the lobster until golden brown and pull out and remove the exceed oil.

❸ 锅中放入黄油、圆葱末煸香，再加入鲜奶油，放入奶酪、盐、白胡椒粉调味，待奶酪融化后放入炸好的龙虾，翻炒均匀，洒入柠檬汁，装盘即可。

Add butter and minced shallot to the pan and swear until fragrant. Then add fresh cream, cheese, salt and white pepper powder to season. After the cheese melts, add the fried lobster and stir-fry evenly. Sprinkle with lemon juice and serve.

技术解析 Technical resolution

在锅内融化奶酪时，要加入足量的鲜奶油，而且要用文火，大火的话奶酪易煳锅底。

When melting cheese in the pot, add enough fresh cream and use low heat. If you use high heat, the cheese will easily stick to the bottom of the pot.

十八、玉米鱼柳
No.18 Fish Fillet with Sweet Corn Sauce

小记 Notes

玉米鱼柳是选用加拿大富产的甜玉米来制作酱汁，配以中式的软炸鱼柳而成。此菜口感咸鲜略带玉米自身的甜味，很受加拿大华人和西方人的喜爱。

Fish fillet with sweet corn sauce is made by using sweet corn, which is abundant in Canada, to make the sauce and paired with Chinese-style soft-fried fish fillets. This dish has a salty and fresh taste with a slight sweetness from the corn itself and is very popular among Chinese Canadians and westerners.

主料 Main materials

龙利鱼。

sole.

配料 Ingredients

玉米粒，鸡蛋。

corn kernels, eggs.

调料 Seasonings

芥花油，黄油，葱姜汁，胡椒粉，淀粉，面粉，盐，鸡粉。

canola oil, butter, onion and ginger juice, pepper powder, starch, flour, salt, chicken powder.

制作过程 Methods

1. 将龙利鱼宰杀去皮，清洗干净切成小长方形条状，用葱姜汁、胡椒粉、盐腌制备用。

   Clean and remove the skin of the sole. Cut into small rectangular strips. Marinate with onion and ginger juice, pepper powder and salt.

2. 碗中放入面粉、淀粉、鸡蛋、水调制成面粉糊。将腌制好的龙利鱼裹匀面粉糊放入油锅炸至金黄，捞出控油。

   In a bowl, mix flour, starch, eggs and water to make a batter. Coat the marinated sole with the batter and deep fried in oil until golden brown. Remove and drain oil.

3. 将一半的玉米粒放入搅拌机内，加入水打碎。另起锅放入黄油，倒入打碎的玉米粒，然后将剩下的玉米粒放入锅中，加入盐、鸡粉调味，加入淀粉勾芡，再均匀地撒入鸡蛋液，放入炸好的鱼柳翻炒均匀，装盘即可。

   Put half of the corn kernels into a blender and add water to blend. In another pot, add butter and pour in the blended corn kernels. Then add the remaining corn kernels to the pot. Seasoning with salt and chicken powder. Thicken with starch. Evenly sprinkle in egg liquid. Add fried fish fillets and stir-fry evenly and serve.

技术解析 Technical resolution

玉米粒不宜打得太碎。蛋花要呈均匀的小薄片。

The corn kernels should not be blended too finely. The whipping egg should be evenly sprinkled into thin slice.

十九、香酥蒜香酱油鸡
No.19 Crispy Garlic Soy Sauce Chicken

小记 Notes

酱油鸡在粤菜中比较常见，以酱油、冰糖、香料等调制卤水卤制鸡肉。而西方人对酱油也是比较偏爱，哪怕吃一碗白米饭，也要在白米饭上淋入酱油。因此香酥蒜香酱油鸡就是在此基础上进行改良的，将酱油作为淋汁浇在炸好的鸡肉上。

Soy sauce chicken is quite common in cantonese cuisine, using soy sauce, sucrose, spices and other seasonings to marinate chicken. Westerners also prefer soy sauce, even if they eat a bowl of white rice, they also need to drip soy sauce on the white rice. Therefore, crispy garlic soy sauce chicken is an improvement on this basis, using soy sauce as a sauce poured on fried chicken.

主料 Main materials

散养鸡。

free-range chicken.

配料 Ingredients

葱段，姜片，蒜末，青葱粒。

green onion section, sliced ginger, minced garlic, chopped green onion.

调料 Seasonings

芥花油，酱油，花雕酒，盐，鸡粉，大料。

canola oil, soy sauce, Huadiao cooking wine, salt, chicken powder, star anise.

制作过程 Methods

❶ 将鸡宰杀清洗干净。

Kill and clean the chicken.

❷ 锅中加入水、花雕酒、葱段、姜片、大料、盐、鸡粉烧开，放入鸡肉，改成小火，将鸡肉煨熟，捞出凉凉，用刷子将鸡身均匀地涂上酱油备用。

Add water, Huadiao cooking wine, green onion section, sliced ginger, star anise, salt and chicken powder to the pot and bring to a boil. Add the chicken and change to low heat. Simmer the chicken until cooked. Remove and cool down. Use a brush to evenly coat the chicken with soy sauce.

❸ 另起锅倒入芥花油烧热，将鸡肉放入油锅炸至表面酥脆金黄，捞出控油。将鸡肉剁成块整齐地摆在盘中。将酱油淋在鸡肉上。

Add canola oil, heat up, deep fry chicken in oil until crispy and golden brown, remove and control oil. Cut the chicken into neat pieces and place them on a plate. The soy sauce drizzled on the chicken.

❹ 锅内留油将蒜末煸至金黄酥脆，连油加蒜末一并倒在鸡肉上面，撒上青葱粒即可。

The remaining oil in the pot will be stir-fried minced garlic crispy gold, oil and minced garlic together with pour in the chicken above, sprinkle chopped green onion can be.

技术解析 Technical resolution

煮鸡肉的卤水盐分不可以太多，否则成菜后再加入酱油，口味会过重。

The salt content of the brine for cooking chicken should not be too much, otherwise adding soy sauce after cooking will be too heavy.

二十、扇贝豆苗蛋花汤

No.20 Scallop Sugar Pea Sprout Egg Soup

小记 Notes

扇贝豆苗蛋花汤是在中餐瑶柱蛋花汤的基础上进行改良制成的，用鲜扇贝丁取代干瑶柱，同时又用西式鱼高汤来增加汤的鲜味。

Scauop sugar pea sprout egg soup is an improvement based on Chinese food dried scallop egg soup, replacing dried scallops with fresh scallop, and using western-style fish stock to increase the umami of the soup.

主料 Main materials

鲜扇贝，豆苗。

fresh scallops, bean seedling.

配料 Ingredients

鸡蛋清，姜丝。

egg white, ginger.

调料 Seasonings

西式鱼高汤，花雕酒，盐，鸡粉，白胡椒粉，淀粉。

western-style fish stock, Huadiao cooking wine, salt, chicken powder, white pepper powder, starch.

制作过程 Methods

❶ 将扇贝清洗干净，加入花雕酒、盐略腌。将豆苗洗净切碎。

Clean the scallops and add Huadiao cooking wine and salt to marinate slightly. Wash and chop the bean seedling.

❷ 锅中加入西式鱼高汤、姜丝烧开，倒入扇贝、豆苗、盐、鸡粉、白胡椒粉调味，再用水淀粉勾芡。

Add western-style fish stock and ginger to the pot and bring to a boil. Pour in scallops, pea sprouts, salt, chicken powder and white pepper powder to season. Then thicken with water starch.

❸ 将鸡蛋清均匀地淋在汤内，成为蛋花。

Pour the egg white evenly into the soup to form an egg slice.

技术解析 Technical resolution

西式鱼高汤的制作方法和中餐完全不同。鱼骨和水烧开之后一定要改成文火，并且要不断地撇去浮沫，然后加入圆葱、西芹、胡萝卜继续文火煮。成品鱼汤为清汤，不是奶汤。

The method of making western-style fish stock is completely different from Chinese cuisine. After boiling the fish bones and water, you must turn down the heat to simmer and constantly skim off the froth. Then continue to simmer with shallot, celery and carrots. The finished fish soup is clear soup instead of milk soup.

西式鱼高汤制作
Making western-style fish stock

原料：鱼骨 2千克，圆葱100克，西芹50克，胡萝卜50克，水4升，白葡萄酒250毫升。

Materials: 2kg of fish bones, 100g of shallot, 50g of celery, 50g of carrots, 4L of water, 250mL of white wine.

做法：将鱼骨剁成块洗净，放入吊汤桶内加入足量的水烧开，改成文火，放入圆葱块、西芹块、胡萝卜块、白葡萄酒煮35分钟，过滤后即可。

Methods: Chop the fish bones into pieces and wash them. Put them in a pot, add enough water to boil and turn down the heat to simmer, add chopped shallot, chopped celery, chopped carrots, white wine and simmer for 35 minutes and strain.

二十一、沙律虾拼盘

No.21 Shrimp Salad Platter

小记 Notes

沙律虾拼盘是典型的将中式烹调方法和西式酱汁融合的菜品。由中餐的盐水虾和酥炸虾饼，再配以哈密瓜、香瓜和西方人日常生活中比较喜爱的美乃滋组合而成。此菜的最大特点：油炸虾饼的油腻因哈密瓜、香瓜和柠檬美乃滋的混搭而变得清脆爽口。

Shrimp salad platter is a fusion of typical Chinese cooking methods and western sauces. It consists of Chinese boiling shrimp and crispy fried shrimp cakes, paired with honeydew, cantaloupe and mayonnaise, which is popular in westerner daily life. The biggest feature of this dish: the greasiness of the fried shrimp cake is turned into a crisp and refreshing taste by the combination of honeydew, cantaloupe and lemon mayonnaise.

主料 Main materials

白虾。

white shrimp.

配料 Ingredients

哈密瓜，香瓜。

honeydew, cantaloupe.

调料 Seasonings

芥花油，葱姜汁，葱段，姜片，花椒粒，胡椒粉，淀粉，鸡蛋，花雕酒，盐，鸡粉，美乃滋，柠檬汁。

canola oil, onion and ginger juice, green onion section, sliced ginger, peppercorns, pepper powder, starch, eggs, Huadiao cooking wine, salt, chicken powder, mayonnaise, lemon juice.

制作过程 Methods

❶ 将一半的白虾去皮、去虾线后洗净，剁成虾泥，加入葱姜汁、花雕酒、鸡蛋、胡椒粉、盐、鸡粉调均匀，做成虾饼，裹匀干淀粉放入油锅炸至金黄，捞出控油。

Peel half of the white shrimp, remove the shrimp thread, then rinse. Chop into shrimp paste, add onion and ginger juice, Huadiao cooking wine, eggs, pepper powder, salt, chicken powder and mix well to make shrimp cakes. Coat with dry starch and fry in oil until golden brown. Remove and drain oil.

❷ 锅中加入水，放入葱段、姜片、花椒粒、盐烧开，将剩下的白虾放入开水中煮熟捞出，放入冰水中过凉，剥皮备用。

Add water to the pot and add green onion section, sliced ginger, peppercorns and salt and bring to boil. Put the remaining shrimp into boiling water and cook until done. Take the shrimp out and put in ice water to cool down. Peel and set aside.

❸ 将哈密瓜、香瓜切成条状，用美乃滋拌匀，铺在盘上，上面摆上剥皮煮熟的白虾，盘子周围摆上炸好的虾饼，将美乃滋加入柠檬汁调成酱汁淋在虾的表面即可。

Mix honeydew and cantaloupe strips with mayonnaise, then place on a plate. Top the plate with the boiled white shrimp, peel the skin, and place the fried shrimp cake around the plate. Add mayonnaise and lemon juice to make a sauce and pour over the shrimp.

技术解析 Technical resolution

虾泥一定要朝着一个方向搅拌上劲。盐水虾在煮熟之后要立刻入冰水，这样剥出的虾脆嫩。

The shrimp paste must be stirred in one direction until it is tight. Boiling shrimp in the saltwater must be immediately put into ice water after cooking so that the peeled shrimp is crunchy and tender.

二十二、爆炒双鲜
No.22 Stir Fried Geoduck and Squid

小记 Notes

爆炒双鲜是借鉴胶东家常菜中的炒海杂拌而来，其烹调方法相同，只是其主料以加拿大盛产的象拔蚌和鱿鱼为主。

Stir fried geoduck and squid is to learn from the Jiaodong home-cooked sea mixed, its cooking method is the same, but its main ingredients to Canada's rich geoduck and squid.

主料 Main materials

象拔蚌，鱿鱼。

geoduck, squid.

配料 Ingredients

芹菜，胡萝卜，白蘑菇，荷兰豆，葱段，姜片，蒜末。

celery, carrots, white mushrooms, snow peas, green onion section, sliced ginger, minced garlic.

调料 Seasonings

色拉油，黄油，花雕酒，白醋，味极鲜酱油，盐，鸡粉，淀粉。

salad oil, butter, Huadiao cooking wine, white vinegar, Weijixian soy sauce, salt, chicken powder, starch.

制作过程 Methods

❶ 将象拔蚌去壳、去内脏清洗干净，片成大片备用。鱿鱼清洗干净，鱿鱼身片成大片。锅内放水烧开，将象拔蚌片、鱿鱼片快速焯水捞出。

Clean the geoduck by removing the shell and viscera and slice it into large pieces. Clean the squid and slice the squid body into large pieces. Boil water in a pot and quickly blanch the geoduck slices and squid slices.

❷ 将芹菜洗净切段，白蘑菇切块，荷兰豆切段，胡萝卜切片，焯水备用。

Wash and cut celery into strips. Cut white mushrooms into pieces. Cut snow peas into segments. Slice carrots and blanch them.

❸ 炒锅中放入色拉油和黄油加热，加入葱段、姜片和蒜末煸香，放入蔬菜、象拔蚌片、鱿鱼片，烹入白醋、花雕酒、味极鲜酱油翻炒均匀，加入盐、鸡粉调味，用水淀粉勾芡，淋入明油出锅即成。

Heat the salad oil and butter in a wok. Add green onion section, sliced ginger and minced garlic to stir-fry until fragrant. Add vegetables, geoduck slices and squid slices. Pour in white vinegar, Huadiao cooking wine and weijixian soy sauce to stir-fry evenly. Season with salt and chicken powder. Thicken with water starch and pour in oil to finish.

技术解析 Technical resolution

象拔蚌焯水和炒制过程中，需要快速完成，时间不能太久，否则象拔蚌的肉质会像胶皮一样，无法咬动。

The geoduck must be blanched and stir-fried quickly. The time cannot be too long, otherwise the texture of the geoduck will become rubbery and difficult to eat.

二十三、山核桃仁什锦蔬菜

No.23 Stir Fried Pecan and Mixed Vegetables

小记 Notes

炒什锦蔬菜也称作炒杂菜，在加拿大的杂碎菜中非常普遍，例如黑豆豉炒杂菜，也可以加入鸡肉、牛肉或猪肉，做成肉类炒杂菜。其烹调方法不变。山核桃仁什锦蔬菜是完全运用中式的烹调技法，在中式炒时蔬的基础上，加入加拿大人比较喜欢的零食“美国山核桃仁”融合而成。

Stir fried mixed vegetables, also known as stir fried miscellaneous vegetables, are very common in Canada’s chop suey. For example, black bean and garlic stir fried mixed vegetables, it can also be added with chicken, beef or pork to make meat stir fried mixed vegetables. The cooking method remains the same. Stir fried pecan and mixed vegetables are completely based on Chinese stir fried vegetables and add pecan that Canadians like.

主料 Main materials

山核桃仁，西芹，玉米笋，荷兰豆，马蹄，胡萝卜，竹笋，白蘑菇。

pecan, celery, corn shoots, snow peas, Chinese water chestnut, carrots, bamboo shoots, white mushrooms.

配料 Ingredients

葱末，姜末，蒜末。

minced scallions, minced ginger, minced garlic.

调料 Seasonings

芥花油，盐，味精，香油。

canola oil, salt, MSG, sesame oil.

制作过程 Methods

1. 锅内放入芥花油烧热，将山核桃仁炸至酥脆捞出控油备用。
   Heat canola oil in a pan and deep fried the pecan until crispy. Remove and drain the oil for later use.
2. 将西芹、玉米笋、荷兰豆切成段，胡萝卜、白蘑菇、竹笋切成块，马蹄切成片。焯水备用。
   Cut celery, corn shoots and snow peas into sections; cut carrots, white mushrooms and bamboo shoots into pieces; cut Chinese water chestnuts into slices. Blanch them in boiling water and set aside.
3. 锅内放入少许油，加葱末、姜末、蒜末煸香，放入蔬菜翻炒，放入炸好的山核桃仁，加盐、味精调味，淋香油出锅装盘即成。
   Add a little oil to the pan and stir-fry minced scallions, minced ginger and minced garlic until fragrant. Add the blanched vegetables and stir-fry. Add the fried pecan and season with salt and MSG. Drizzle with sesame oil and place on the plate.

技术解析 Technical resolution

炸核桃仁酥脆上色即可，其他食材焯水，用于除去异味。

Fried pecan until light color and crispy color, other ingredients blanch in the water, to remove odor-based.

二十四、爆炒鸡软骨
No.24 Stir Fired Chicken Cartilage

小记 Notes

鸡软骨的应用在北美传统西餐中是比较少见的，但在中西融合的中餐杂碎菜中比较常见。运用西餐的烤制法再加以中餐的烹调方法融合而成。

The use of chicken cartilage is relatively rare in traditional western cuisine in North America, but it is quite common in the fusion of Chinese and western cuisine. It is a fusion of western roasting methods and Chinese cooking methods.

主料 Main materials

鸡软骨。

chicken cartilage.

配料 Ingredients

圆葱条，青红椒条，蒜末。

shredded shallot, shredded green and red pepper, minced garlic.

调料 Seasonings

橄榄油，芥花油，酱油，糖，盐，味精，黑胡椒粉，白芝麻，面粉，湿淀粉，汤汁。

olive oil, canola oil, soy sauce, sugar, salt, MSG, black pepper powder, white sesame , flour, wet starch, soup.

制作过程 Methods

❶ 将鸡软骨清洗干净，拌入盐、黑胡椒粉腌制，然后裹匀干面粉。均匀排在烤盘内，在鸡软骨表面刷上一层橄榄油，用锡纸裹住烤盘，放入220℃烤箱内烤10 ~ 15分钟。

Clean the chicken cartilage and marinate with salt and black pepper powder. Then coat with dry flour. Arrange evenly on a baking sheet. Brush a layer of olive oil on the surface of the chicken cartilage. Wrap the baking sheet with foil and bake in a 220℃ oven for 10 ~ 15 minutes.

❷ 另起锅倒入芥花油烧热，放入蒜末、圆葱条、青红椒条炒出香，加入酱油、糖、味精和少许汤汁，放入烤好的鸡软骨翻炒均匀，用湿淀粉勾芡后装盘，撒上白芝麻。

In the wok, heat canola oil and add minced garlic, shredded shallot, shredded green and red pepper, to stir-fry until fragrant. Add soy sauce, sugar, MSG and a little soup. Then place baked chicken cartilage and stir-fry evenly. Thicken with wet starch and sprinkle with white sesame .

技术解析 Technical resolution

由于每个烤箱的温度不相同，所以在烤制前需要测试，要控制好鸡软骨的烤制时间，以免出现焦煳或者断生。

Since the temperature of each oven is not the same, it is necessary to test before roasting to control the roasting time of the chicken cartilage to avoid burning or undercooking.

二十五、豉椒鸡片

No.25 Chicken Slices with Black Bean Sauce

小记 Notes

黑豆豉在北美是家喻户晓的中餐调料之一。西方人在中餐馆就餐时常点带有豆豉调料的菜品。

Black bean paste is one of the well-known Chinese seasonings in North America. Dish with bean paste seasoning that westerners often order when dining in Chinese restaurants.

主料 Main materials

鸡胸肉。

chicken breast.

配料 Ingredients

圆葱，青椒，红椒，白蘑菇。

shallot, green pepper, red pepper, white mushroom.

调料 Seasonings

芥花油，姜末，蒜末，料酒，酱油，黑豆豉，盐，味精，淀粉。

canola oil, minced ginger, minced garlic, cooking wine, soy sauce, black bean paste, salt, MSG, starch.

制作过程 Methods

❶ 将鸡胸肉切成薄片，放入盆中用料酒、盐、淀粉腌制。

Cut the chicken breast into thin slices and marinate with cooking wine, salt and starch in a bowl.

❷ 将圆葱、青椒、红椒、白蘑菇切成小块备用。

Cut shallot, green pepper, red pepper and white mushroom into small pieces and set aside.

❸ 锅内放水烧开，将鸡胸肉片焯水捞出备用。青椒、红椒和白蘑菇焯水备用。

Boil water in a pot. Blanch the sliced chicken breast and set aside. Blanch green pepper, red pepper and white mushroom and set aside.

❹ 净锅倒入芥花油加热，煸香圆葱、姜末和蒜末，放入黑豆豉、酱油和焯好水的鸡胸肉片、青椒、红椒、白蘑菇，用盐、味精调味翻炒均匀，出锅即可。

Heat canola oil in a wok, add shallot, minced ginger and minced garlic to sweat until fragrant, then add black bean paste, soy sauce and blanched chicken slices to mix well. Add green pepper and red pepper and white mushroom. Season with salt and MSG and stir-fry evenly and serve.

技术解析 Technical resolution

鸡胸肉切片要薄，焯水要快速，以保持鸡片的嫩滑。

Chicken breast slices should be thin, and blanching should be done quickly, to keep the chicken slices tender smooth.

特别荣誉菜

Special Honor Dishes

国汤手拉冲浪活海参

Chinese Signature Soup Poached Hand-stretched Live Sea Cucumber

小记 Notes

此菜是最具代表性的“中餐分子美食”，资深级注册中国烹饪大师高速建先生，独立发现了活海参的两个“适口点”（最佳口感），从而发明了“中餐低温慢煮”的分子美食技法菜肴——手拉冲浪活海参。同时填补了中餐海参烹饪食材的种类（手拉活海参）和海参烹饪原料的涨发加工方法（手拉法）的空白。

This dish is the most representative “Chinese molecular cuisine” created by Mr. Gao SuJian, registered Chinese culinary master, who found the two “best cooking points” that bring out the best taste of live sea cucumber, and invented his signature molecular dish—Poached Hand-stretched live sea cucumber with sous vide method. At the same time, this dish fills in the variety of Chinese sea cucumber cooking ingredients (Hand-stretched live sea cucumber) and the swell process of sea cucumber cooking ingredients (Hand-stretched technique).

1996年，由世界中餐业联合会主办的第二届中国烹饪世界大赛中，高速建大师制作的“国汤烩乌鱼蛋”，对传统“烩乌鱼蛋”单一用醋与国宴“烩乌鱼蛋”只用酸黄瓜汁的调味方法加以优化创新，采用“优选法”将醋和酸黄瓜汁按一定比例调和，并勾薄芡，使其口味更加酸辣柔和、口感更加爽滑适口，妙不可言，得到了国内外评委的高度评价而荣获头名状元的金牌，从而升华了“国汤”的品位，被业内誉为“国汤”的第三部曲（第三版本）。

In 1996, in the second Chinese Culinary World Competition sponsored by the World Federation of Chinese Catering Industry, Master Mr. Gao SuJian optimized the traditional “Chinese siqnature soup braised cuttlefish roe soup” with vinegar as well as the signature soup of state banquet - “braised cuttlefish egg soup” only with the sour cucumber juice, and mix vinegar and sour cucumber juice in a certain proportion, and add a small amount of starch to thicken its taste more sour, spicy, soft, smooth and more palatable. It was highly praised by domestic and foreign judges and won the gold medal of the first champion, thus sublimating the taste and was also known in the industry as the third version of “Chinese signature soup”.

此菜集大世界基尼斯纪录、国家发明专利及中餐分子美食（低温慢煮）于一身，融中西饮食文化、技艺传承创新、书画艺术、科学养生为一体，将中西方饮食文化合璧理念演绎到极致，活海参通过手拉技法，将海参拉到吉尼斯薄度（0.003毫米以内），大幅度提高了海参的吸收率，加以用“国汤”口味的酸辣汤汁冲制，让客人感受（体验）到饮食文化与书画艺术的完美结合，从而让客人不但一饱口福，而且一饱眼福和耳福——品海参美食、听海参故事、观海参绝技、赏海参膜画。

This dish integrates Great World Dsjjns Record, national invention patents and Chinese molecular cuisine (low temperature slow cooking method), Chinese and western food culture, culinary technology inheritance and innovation, calligraphy and painting art, scientific health preservation as one, interpreting the concept of combining Chinese and western food culture to the extreme.Through hand-stretched technique, the sea cucumber is pulled to the thinness of guinness (within 0.003mm), and the absorption rate of sea cucumbers has been greatly improved. The outstanding sour and spicy taste of the "Chinese signature soup" is also used in this dish, allowing guests to experience the perfect combination of food culture and calligraphy and painting art to enhance their dining experience from visual and auditory—taste the sea cucumber, listen to its stories, and enjoy its gorgeous presentation.

上海大世界基尼斯總部

大世界基尼斯之最

最薄的“手拉活海参”

平均厚度：0.019毫米

高速建（北京）将活海参（约重75克、直径3厘米、长10厘米），经过2分钟的抻拉，达到长70厘米、宽45厘米、薄0.002毫米的完整薄片。

NO：02166
2008.12

王以幸　　蔡丰

手拉活海参基尼斯纪录证书

国家发明专利作品——海参膜字画

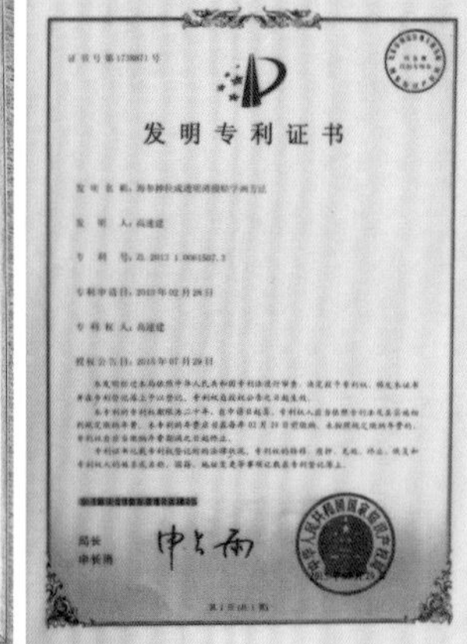

发明专利证书

“海参膜字画”发明专利证书

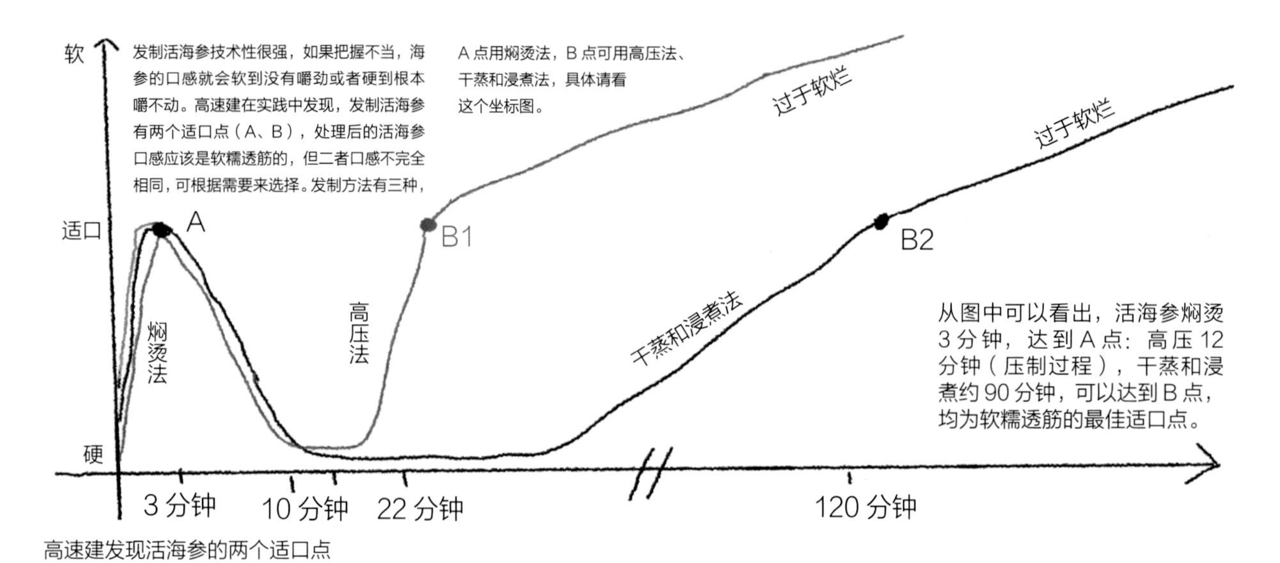

高速建发现活海参的两个适口点

高速建表演手拉活海参

主料 Main materials

活海参。

live sea cucumber.

配料 Ingredients

竹荪段，海苔，枸杞，酸黄瓜末，香葱末，香菜末。

bamboo fungus section, nori, wolfberry, minced pickle cucumber, minced scallion, minced cilantro.

调料 Seasonings

鱼露，御制酸汤汁，酸黄瓜汁，白醋，盐，花雕酒，白胡椒粉，高汤。

fish sauce, royal fermented rice sauce, pickled cucumber juice, white vinegar, salt, Huadiao cooking wine, white pepper powder, stock.

制作过程 Methods

1. 将活海参拉伸成基尼斯纪录的完整大薄片（0.003毫米以内），然后用剪刀将其剪成连刀长条，放入50～55℃的水中浸泡30分钟，捞出盛入汤盅内。

   Stretch the live sea cucumber into a complete piece of Dsjjns record (within 0.003mm), then cut it into a connecting strip with scissors, soak it in 50 ~ 55℃ water for 30 minutes, remove and put into soup bowl.

2. 将泡好的竹荪段、枸杞分别用沸水焯煮一下，捞出放入盛海参的汤盅，然后放入剪成小菱形的海苔、酸黄瓜末、香葱末、香菜末。

   Blanch bamboo fungus section and wolfberry with boiling water, remove and put into the soup bowl of sea cucumber, then add the nori, minced pickled cucumber, minced scallion and minced cilantro.

3. 锅中放入高汤烧开，加入花雕酒、鱼露、盐、御制酸汤汁、酸黄瓜汁、白醋、白胡椒粉调好口味，撇去浮沫，盛入汤壶中。

   Put stock in a pot and bring to a boil, Add Huadiao cooking wine, fish sauce, salt, royal fermented rice sauce, pickled cucumber juice, white vinegar, white pepper powder to adjust the taste, skim the foam, into the soup pot.

4. 汤盅和汤壶一起上桌，将汤壶中调好口味的高汤冲入汤盅即可食用。

   Serve the soup bowl and pot together, pour the stock from the pot into the soup bowl and serve.

技术解析 Technical resolution

冲制的高汤温度不要超过80℃，以免影响海参的口感。

The temperature of the stock should not be higher than 80℃, otherwise the sea cucumber will easily overcook.

后记

在本书的文稿付梓之际，与其他编书之人一样不能免俗，总觉得有几句话需要在最后絮叨一下，敬请各位读者朋友见谅!

关于“中西餐合璧菜”的话题，早在20世纪80年代末就被提出来了。中国鲁菜特级大师、烹饪泰斗王义均的高足高速建先生，曾于1989年在山东省烹饪学会会刊《烹饪者之友》杂志第三期上，发表了《中西合璧菜肴》的短文章，引起了业界的关注。及其后，高速建先生又于1993年在《烹饪者之友》再次发表了《融中西烹技，制人间至味》的文章，对中西餐合璧菜的概念、特点进行了论述，并就中西餐合璧菜研究发展的必要性以及渊源进行了初步的探讨。这篇文章指出，中西餐合璧菜肴的实践案例早在20世纪80年代初就已经出现了，发表于1985年第7期《中国烹饪》的“空心黄油虾球”就是一款中西餐结合的菜品，它把中式烹饪中的“空心虾球”和西式烹饪中的“黄油虾卷”融为一体，使菜品的风格令人耳目一新。

然而，时光荏苒，蹉跎间30多年过去了，关于中西餐合璧菜肴的实践成果已经在我国广泛普及开来，但理论方面的研究成果却几乎依然停留在往昔之中。迄今为止，几乎没有一本专门研究中西餐合璧菜肴的理论著作，甚至连一本像样的记录中西餐合璧菜肴烹饪技艺的菜谱都没有面世。虽然也有中国香港万里机构出版的《中西合璧套餐》之类的小册子，但无法反映出近几十年来中国厨师在中西餐合璧菜肴的研究与实践中所做出的成果，令人唏嘘不已!

本书精选了56款具有中国、加拿大两国融合风格的菜肴，并通过中英文双语的诠释，向国内外的烹饪工作者、爱好者及广大读者，提供一本久违了的难得好书。言赘必烦，敬请各位朋友亲自阅读、品鉴。

本书的出版，得到了世界中餐业联合会国际中餐名厨专业委员会、加拿大餐饮总会和加拿大中华烹饪协会的鼎力支持，世界中餐业联合会会长邢颖先生、加拿大餐饮总会会长黄汝週先生、加拿大联邦烹饪协会会长瑞安·马奎斯和加拿大中华烹饪协会会长廖志文先生分别为本书写了序，在此表示衷心的感谢!

最后，真诚地感谢为此书的撰写、编辑、编审辛勤劳动以及关心《中西餐合璧菜：中国和加拿大》一书出版的所有朋友!

编者谨记

2023年10月1日于济南